NAUTILUS

NAUTILUS

The Story of Man Under the Sea

ROY DAVIES

BBC BOOKS

Dedicated to the memory
of Paul Johnstone
who showed so many of us
where the road began.

TITLE PAGE PHOTOGRAPH
Control room of a nuclear-powered fast attack submarine, USS *Olympia*.

This book is published to accompany the television series
entitled *Nautilus*: which was first broadcast in 1995.
The series was produced by BBC Wales.
Executive Producer: Roy Davies
Series Producer: Anita Lowenstein
Directors: Ian Potts, Anita Lowenstein and Jeremy Marre

ISBN 0 563 37043 2
Picture research by Anne-Marie Ehrlich
Map by Mike Gilkes
First published 1995
Published by BBC Books
an imprint of BBC Worldwide Publishing
BBC Worldwide Limited,
Woodlands, 80 Wood Lane, London W12 0TT

Set in Joanna by Ace Filmsetting Limited, Frome
Colour separations by Radstock Reproductions, Midsomer Norton
Printed and bound in Great Britain by Cambus Litho Ltd, East Kilbride
Jacket printed by Belmont Press Limited, Northampton

CONTENTS

INTRODUCTION

This is a story about man's determination to develop machines from which he could fight and that would protect him while he explored the incredible world which begins at the surface of the sea.

The figures which tell that story can be found in many books. Details of submarine strategy, its effects, its limitations, its successes and failures are now generally accepted. The depths of the ocean, the pressure of water, a thickness of steel, the shape of a hull. This country destroyed so many million tons of shipping; that campaign resulted in so many submarines being sunk. But the figures are usually dead. Stark. Fixed. They record, but they tell no human story. They allow comparisons to be made about relative success and failure but, as indicators of human frustration, or despair or boredom, they have little meaning and almost no value.

It is only when we begin to hear men talk of the conditions under which they lived, of hardship and determination, that printed figures begin to breathe. It is only when those who were there describe the bravery, the comradeship and the sacrifice they experienced that lines of print begin to move and stretch and take the shape of men. Then, and only then, do those stark details, which are supposed to represent a collective history, have any relevance to us today.

This, then, is a book about just such men and about the machines inside which they lived, explored, fought and sometimes died. A few men, now themselves dead, have left us firsthand accounts. Some, still alive, wanted us to know how it really was. Were it not for their honesty, neither this book, nor the television series it accompanies, could ever have been contemplated.

Nor could such an account have been achieved without the generous help of several colleagues. I want to acknowledge the help and also advice given to me

USS *San Francisco*, a Los Angeles class nuclear-powered attack submarine, leaving Pearl Harbor channel for deployment in the Pacific.

by Norman Polmar, whose books have proved constant reference points; Victoria A. Kaharl, who has generously allowed me to use a section from her book *Water Baby*, published by Oxford University Press, which evokes one of the most exciting moments in underwater detective work; Mike Dash, Associate Publisher of *Viz* magazine, whose unpublished thesis on the origins of the modern submarine captures so much of the political motivation of the times and who read and corrected my attempts to chart a way through that period of the story; Nicholas Lambert who delivered invaluable information and guidance about the complexities of national attitudes and strategies during the First World War; and Richard Compton-Hall, a writer who has done more than most to introduce the submariner's world to our own, for reading the typescript with the critical and practised eye of both submarine skipper and historian. I continue to be amazed that writers and academics of such standing can put aside their own work at extremely short notice to help out colleagues of whom they can hardly have heard.

Of those who did know me and still wanted to offer their help, I must record my gratitude to Martha Caute, my editor at BBC Books, for her support and insight, Anita Lowenstein, my series producer, for her energy, attention to detail and encouragement, and Ian Potts, Jeremy Marre, Jonathan Hacker, Kate Parry, Tessa Coombes, Gabriella Romano, Steve Bergson, Daniella Mamo and Diane Bartley, for their constant goodwill throughout the project. I want, also, to thank my colleagues at BBC Wales and especially Anwen Davies for bearing, with such good humour, my stories of submarines and submariners to which they have found themselves subjected for the past twelve months.

Inevitably, such a project has not left much spare time for my wife, Marilyn. I can only say that without her selflessness and understanding, it would have proved impossible for me to have accepted this commission. She has acted as psychologist, historical adviser, enthusiast, audience, editor and perceptive critic throughout my career. It has been no different with this project. I am a very lucky man.

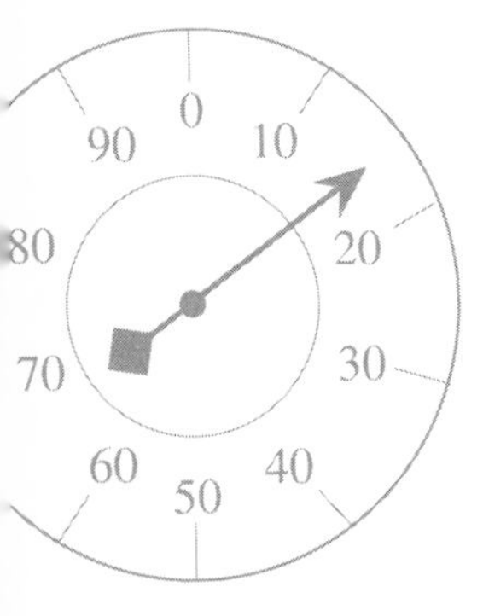

Finally, despite all the care and checking, mistakes of one kind or another might have crept in. It goes without saying that I claim them all. I only hope they are not so glaring as to ruin your enjoyment of the story which follows.

Roy Davies
Pontycymmer
February 1994

'The sea does not belong to despots. Upon its surface
men can still exercise unjust laws, fight, tear one another to pieces, and
be carried away with terrestrial horrors. But at thirty feet below its level,
their reign ceases, their influence is quenched and their power disappears.
Ah! sir, live – live in the bosom of the waters! There only is independence!
There I recognize no masters! There I am free.'

Jules Verne: *Twenty Thousand Leagues Under the Sea*

PROLOGUE

The entire crew of USS *Nautilus* held its breath. The trace on the graph which indicated how far the ice reached down had dropped even further towards the line indicating the top of their 'sail', the huge tower that is so distinctive a part of the submarine's design. They were now only 20 ft (6 m) above the sea-bed and the ice was about to squeeze them into less and less water.

Captain William R. Anderson, called to the attack centre from his cabin, was told that they had just gone under an ice ridge which reached down 63 ft (19 m) from the surface. *Nautilus* had scraped under by only 8 ft (2.5 m). He immediately reduced speed to dead slow. The submarine crept forward between the sea-bottom and the ice.

Captain Anderson, searching, probing, sensing for a passage under the polar ice cap in June 1958, was relieved they had negotiated a way past that first obstacle. 'But we were still in trouble. Our instruments told us an even more formidable barrier lay just ahead. I stared in disbelief at its picture on the sonar. The books said this couldn't happen.

'Slowly, very slowly, we moved forward. My eyes were glued to the recording pen. Downward it swooped again – down, down, down. I reflexed, as if to pull my head into my shoulders. How I wished I could do the same with *Nautilus*.'

In his mind Anderson likened the situation to that of a small boy trying to squirm beneath a fence under which he might well get stuck. 'The inevitable consequences could be severe damage to our ship, perhaps even slow death for those on board. I waited for and honestly expected the shudder and jar of steel against solid ice. The recording pen was so close to the reference line which indicated the top of our sail that they were, for what seemed like hours, almost

USS *Nautilus*, her lines reminiscent of German submarines of the Second World War, approaches full surface speed during sea trials in 1975.

The deck officer of *Nautilus* on alert as the submarine fringes the polar ice cap on its way to the Arctic ocean.

one and the same. I, and others in the attack centre, I am certain, turned for assistance to the only person who could help us.

'In pure agony we stood rigidly at our stations. No man moved or spoke. Then, suddenly, the pen which had been virtually stationary, slowly moved upward. The gap between the ice and *Nautilus* was widening. We had made it. We had cleared, by an incredible 5ft, a mass of ice big enough to supply a 100 lb block to every man, woman and child in the USA.

'It took only a second's reflection for me to realize that Operation Sunshine had totally and irrevocably failed. Not even *Nautilus* could fight that kind of ice and hope to win. To the north of us lay many miles of even shallower water and possibly even deeper ice. There was no question about it. The only sane course was south.'

America's first attempt to take a nuclear-powered submarine from the Pacific to the Atlantic by way of the polar ice cap was over.

But three months later, in slightly warmer conditions and by a slightly different route, *Nautilus* again nosed under the ice cap, found a deep, uncharted channel, and sped for the Pole without interruption. The submarine crossed 90°N at 23.15 (11.15 p.m. Eastern Daylight Saving Time) on 3 August 1958 and emerged at the Greenland Spitsbergen portal a day and a half later.

For the first time a submarine, creating its own oxygen and with immense reserves of power, had navigated the last, unexplored corner of the inner universe. The achievement was universally recognized as incredible. But few realized that *Nautilus* that day dragged in its wake larger and far deadlier shadows which were to haunt the world for half a lifetime.

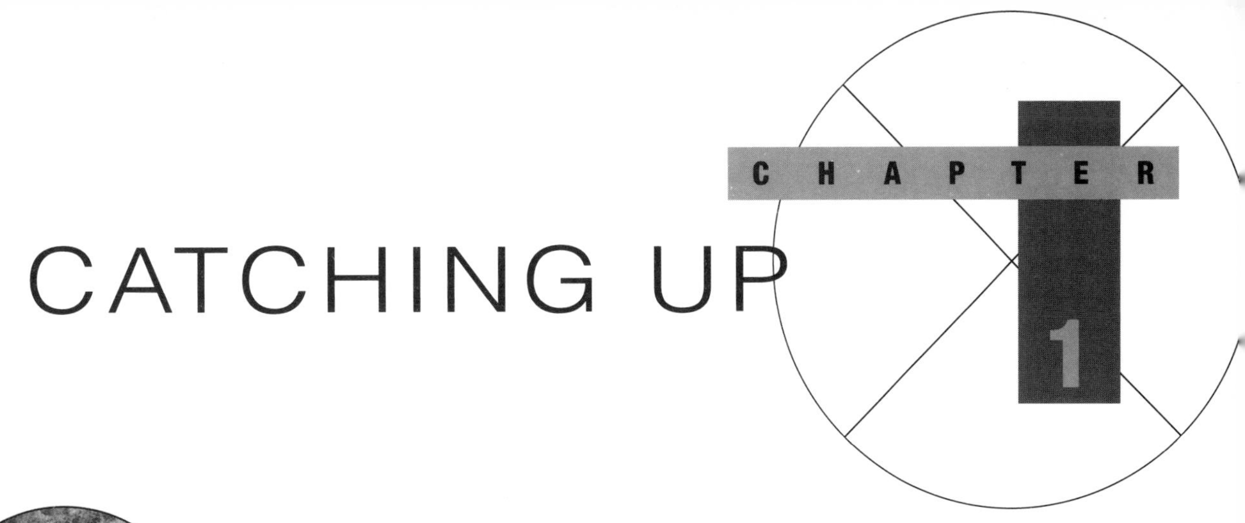

CHAPTER 1

CATCHING UP

One day in the late autumn of 1952, Engineer-Captain 1st Rank Vladimir Nikolayevich Peregoudov, the most accomplished submarine designer in the Soviet Union, crossed Red Square on his way to the Kremlin. He had been given no indication as to why he had been summoned, but the man he was to see, Viatcheslav Malychev, Vice-President of the Council of Ministers of the USSR, was one of the most powerful men in Stalin's Russia at a time when the Cold War was approaching its most intense period. The meeting was to begin one of the most incredible stories of our time.

Malychev came directly to the point. He told Peregoudov that Russian diesel submarines were no longer up to standard. They could not even match the best German submarines of the Second World War, which had ended seven years before. Their batteries were second-rate and the speed and depth at which they could operate were no longer acceptable. State Security, the KGB, had discovered that the Americans were soon to launch a submarine that would be powered by atomic energy and Stalin was determined that the Soviet Union should not be left behind. A nuclear submarine would travel at three times the speed of conventional Soviet ones and could stay under the sea for months without needing to surface. The military implications were obvious: if the Soviet Union did not act immediately, the Americans would regain the superiority they had paraded until the Soviet Union had exploded its own atomic bomb three years previously. 'I'm not trying to flatter you,' Malychev told Peregoudov, 'but I see no one else who could direct this project.'

'There has never in peacetime been anything comparable to the current growth in Russian Naval power', admitted Admiral Hyman G. Rickover shortly before the Americans learned of the existence of the USSR's Typhoon submarine, the largest ever built, which displaced 26 500 tons and was more than 558 ft (170 m) long.

Stalin's decision to order the production of a nuclear submarine was very recent. He had been briefed on the idea only weeks before by the atomic physicist, Academician Anatoli Alexandrov. The top secret meeting had been arranged when Stalin read a letter Alexandrov and Igor Kourtechev, the father of the Soviet atomic bomb, had written some months previously suggesting a research project into the feasibility of a nuclear-powered submarine.

Pavel Kotov, now Admiral Kotov, was assigned to work on the project as a young naval officer, and described the secrecy that surrounded it. 'It was so secret that a lot of people among the naval authorities didn't even know about the existence of such a project, and the ministry to which it was allocated was not the Ministry of Defence, but the Ministry of Medium Machine Building... It was so secret that even Bulganin [then Minister for Defence] had to ask one of his contacts to give him access to the top secret documents because he couldn't get them through the Ministry of Defence.'

Stalin was desperately concerned about the American initiative. Apart from the scientific community that had given the Soviet Union its atomic and then its hydrogen bombs, few people could make any meaningful contribution to the creation of a nuclear submarine. There was no widespread knowledge or teaching of nuclear physics and engineering in Russia as there was in the West. Nor were the basic facts about nuclear physics available in reference or textbooks, so paranoid was Stalin about secrecy.

The time allocated to complete the project was very short and everything the Soviet economy and industry could do or provide to ensure the success of the project was made top priority. Although the scientists and engineers were given ideal conditions in which to work, the mental and physical requirements and demands made on them were tough.

Stalin had decided to give the project to the Ministry of Medium Machine Building, one of the most secret Soviet departments, rather than to the Department of the Navy, simply because he was convinced that the latter could no longer be trusted to keep his secrets. After the war, Admiral Nikolai Kuznetsov, Commander-in-Chief of the navy, had been tried, with his three senior deputies, for the 'treason' of transferring a German torpedo design to the British. Kuznetsov was exiled to the Far East fleet as a Rear-Admiral for three years and his deputies were given prison sentences.

A K class Russian diesel submarine of 1942. Ten years later its successors were still technically inferior to the German boats of the Second World War.

As Peregoudov's meeting with Malychev moved to a close, he realized he had little choice. If Stalin wanted him to work on the project, he was going to have to work on it. He was told that he would be working closely with two academicians: Nikolai Dollejol, who would be in charge of designing the reactor for the submarine, and Guenrikh Gassanov, who would design and construct the steam generators.The challenge of building a nuclear submarine captured Peregoudov's imagination, but he pointed out to Malychev that he was over fifty years of age and his health had suffered during a spell in one of Stalin's prisons. Malychev, with his own problems, did not want to hear reasons why Peregoudov might not be able to complete the task. He told him not to lose a minute in getting started or they would both pay dearly for it. Stalin had told Malychev that he wanted the project completed in two years. Did Peregoudov understand?

The designer was all too aware of the unspoken warning behind the question. The son of a peasant and a brilliant mathematician and engineer, as a young man he had worked with Boris Malinin and Mikhail Roudnitski, the men who created the first Soviet submarines. Years later, when he was renowned for the audacity of his technical solutions to problems, Peregoudov was given the job of constructing the Type 'S' submarine which the Soviet Union used in the Baltic in the Second World War.

When Sergei Tourkov, the director of the Type 'S' programme, displeased Stalin and was sentenced to imprisonment, Peregoudov, who had known him since the age of eleven, was instructed to condemn his friend as a traitor and an enemy of the people. He refused – and continued to refuse despite torture and captivity. Eventually he was released and resumed his work. Sergei Tourkov never left Stalin's prison.

Within a few days Peregoudov, Dollejol and Gassanov had started talking and planning in locked rooms. They very quickly discovered that each of them had only a limited understanding of the kind of vessel they had been ordered to create. Dollejol had never even seen a submarine let alone what might be inside one; Peregoudov knew nothing about nuclear physics or the kind of plant that would be required to drive such a boat. The entire Soviet nuclear submarine programme had been given to three men not one of whom had a clear, comprehensive idea of the range of problems they were about to face.

Each man had to familiarize himself with his colleagues' area of expertise. Peregoudov visited the Obninskoie civil nuclear station to study the reactor plant and the steam generators that drove the turbines which produced electricity for Soviet industry. Working and watching every hour possible, he questioned the men in charge about their work, their responsibilities and the nature of the reactor itself.

His colleagues visited submarines in naval bases and talked to the officers and men who lived, worked and fought in the narrow, confined and dingy compartments that served as their living rooms and workplaces. Dollejol admitted on his return that he had been terrified by the lack of discipline in the way the men lived on board the submarines. He told Peregoudov that he did not know how they were able to live like that and was told that submariners were a different breed: 'We'll offer them better conditions on board our submarine,' Peregoudov promised.

From morning to night over the weeks which followed the three men and the teams they had recruited calculated, designed and re-calculated sizes and shapes of compartments, of turbines, of generators, of reactors and pipework and batteries and shafts and sleeping areas and eating areas and torpedo tubes and the configuration of the control centre and its equipment and a thousand other details. Vladimir Barantsev, one of the men drafted in to work on designing the transmission systems of the submarine, remembered the intense secrecy surrounding the project: 'I was never officially told what we were designing. I had to guess it myself. I didn't know what we were engaged in. The word reactor was never pronounced out loud. It was called a crystallizer not a reactor and it took me about three or even four months to understand what kind of submarine we were designing. Each person was pursuing his own business. I was involved with the mechanical plant.

'I was assigned to make calculations of the speed of the submarine but I realized that the driving force on which I was told my calculations had to be based could not be delivered by ordinary batteries. When I looked at the drawings and saw the dimensions of the shafts I guessed everything but I was advised not to discuss it with others.

'There were also difficulties in learning to cope with and master the new technology. This was partly because of the secrecy that surrounded the project – many of the enterprises with whom orders were placed for this or that part of the technology didn't know why or for what they were designing this or that piece of equipment.'

Peregoudov was responsible for designing, from the shape of its hull down to the smallest internal detail, a submarine that would work as a fast, safe, deep-running and menacing weapon of war.

When its designers faced problems they went straight to him. When the nuclear scientists faced problems they went to Alexandrov, who had been given scientific control of the project.

The scientists referred to Alexandrov for all kinds of advice on the project which they all knew as the K-3. How would a nuclear reactor behave in a boat at sea rather than on solid foundations on land? Would the nuclear pile be able to stand the rolling and pitching of a submarine on the surface of the sea? Would a reactor operate efficiently in the strict confinement of a submarine running deeply submerged?

Alexandrov must have been thrown by some of these problems but he could not escape the most crucial question: what kind of reactor was to be designed for the submarine? Whatever the eventual decision, it would still only be a heat source to produce steam. But there were long discussions over what should constitute its core; what method should be used to vary the amounts of nuclear activity in order to control the levels of heat available to make the steam; what method should be used to conduct the heat to the water to produce steam; and how thick the radiation shields would have to be to protect the crew from gamma rays. Every question had several alternative answers.

The basic problem was to calculate the weight and dimensions of the nuclear reactor and plant needed to drive a submarine of a given tonnage. In Alexandrov's words: 'Without this information Peregoudov, the chief designer of the submarine, lacked the essential information he needed to design the boat itself. From the start everybody realized that the size of the reactor would determine the diameter of the submarine.' Once Peregoudov had indicated that the optimum diameter of the boat he wanted to build would be 33 ft (10 m) he informed the scientists that the atomic reactor could not occupy more than one third of that diameter, because he intended using two reactors side by side to drive the submarine.

This was an important breakthrough but not one which solved the greatest problem. Alexandrov again: 'Our first requirement was to make each reactor powerful enough. We wanted the submarine to develop a speed of 30 knots and so the reactor systems had to produce enough energy to drive the turbines to enable the boat to develop that speed. But the reactor is a complicated system and cannot be smaller than a certain size because it has to accommodate a lot of elements. If these were not included the reactor couldn't be guaranteed to produce the required amount of power. But it couldn't be bigger than the submarine could accommodate. Faced with such a problem we scientists found ourselves between the devil and the deep blue sea.'

A nuclear reactor is not a potential nuclear bomb but a device designed to permit a succession of controlled and self-sustaining collisions between neutrons and atoms in order to generate energy and hence heat to produce steam which drives the turbines. The ability to control the amount of steam produced allows the captain to vary the speed of the submarine. 'Our first priority was to make everything reliable,' Alexandrov said. 'But everything had to be improvized. The scientists had to decide on the spot whether this or that was better, and everyone was asking the question why this and why not that. In the event a brilliant, absolutely brilliant, reactor was designed which was 1.5 metres in diameter and whose walls were 2 centimetres thick. The lid was about 4 centimetres thick.'

By the autumn of 1953, only four months before the Americans planned to launch their first nuclear submarine, the Soviet scientists were in a position to test their reactor for the first time. 'I remember there was a lot of concern and worry when the nuclear plant was switched on because it was all happening for the first time,' said Vladimir Barantsev. 'It was a complete novelty; something where the consequences were totally unforeseen. Everything was done in great secrecy.' Admiral Kotov also recalled the excitement of the occasion: 'When the K-3 reactor was tested, it indicated a speed of 30 knots – a speed which would go on endlessly. It could drive a ship which wouldn't need to surface for a long time. We knew it could reach not only England, it could reach America without surfacing and come back.'

But the immensity of the task faced by the submarine's designers only became clear during 1953 when they also faced other less exciting but equally important problems. In Barantsev's words: 'It became clear to us what a large-scale task it was – not only the nuclear engineering but also the materials in which the boat should be constructed. We needed new construction materials, new acoustic systems, new habitation systems and many more innovations because the nuclear submarine was going to be much more self-contained, a more autonomous, submarine, and its velocity was going to be much higher than that of conventional submarines. Remote control systems were to be introduced into this boat to a much larger extent than in any previous boat.'

Admiral Kotov remembered that because everything was experimental there were problems in ensuring standards of quality. 'In the course of designing and construction, all of a sudden one had to replace materials or replace entire structures. First, a model was built and then a simulator imitating several compartments of the submarine. And when we started testing, we found that one thing fitted, another thing didn't. For example, the hull had to be increased in size. Stainless steel was used for pipes. But the pipes were subject to 200 atmospheres of pressure and cracks appeared in the steel, even when the water they were carrying was purer than the distilled water which pharmacists sell. Things like that happened and you had to solve these problems very quickly when either the material or the design proved wrong.' The scientists and engineers working on the technology of the boat solved difficulties as they went along, not really knowing where or when the next problem would present itself.

As they did so, Peregoudov strove for an entirely new shape for the hull of the nuclear submarine, which would allow it to be close to hydrodynamically perfect when travelling submerged. Under his influence, designers and draughtsmen began to create an entirely new concept of a true submarine: a boat that could spend an indefinite time under water without needing to emerge for air.

Barantsev remembered the problems Peregoudov faced: 'As in every big project there were completely contradicting designing requirements. For example, the boat had to be as silent as possible. Therefore, it had to be smooth and perfectly shaped and couldn't afford to have things sticking out. Also, it had to be able to withstand the pressure of ice because the submarine would travel into the northern seas. Yet it had to be designed so that it had all-round sound reception with no other constructions which might block the perfect view. These were the kinds of demands that Peregoudov and the other designers had to consider.'

Finally, having considered every possible outline that was wider in circumference at the bow than towards the stern, Peregoudov settled for the shape of a whale and led his designers into the task of creating the form in steel. Alexandrov remembered: 'Peregoudov wasn't wedded to familiar shapes. He looked for new shapes and made those work.'

When he first saw Peregoudov's design for the K-3, Barantsev both admired the revolutionary shape and realized the headaches it was going to cause the construction teams: 'The shipyard people were very angry when they realized they had to build a submarine of such a shape. It was a very beautiful shape but very complicated to build. The layers of material which comprised the hull were so thick, and they knew how difficult it was going to be to give such a shape to a submarine when you had to build it with material of such thickness.'

When the final blueprints for the K-3 were approved in 1954, the shape of the hull and Peregoudov's plan to incorporate twin nuclear reactors inside it were immediately seen as impressive innovations. But no one could have guessed at the success these features were to bring the Soviet Union in the submarine race with the Americans that was about to start.

Two years after Peregoudov walked back across Red Square after his historic meeting with Viatcheslav Malychev, Lev Zviltsov, a submarine commander serving with the Soviet Black Sea fleet, passed the guards on his way into the Kremlin. A strong, upright man, he had absolutely no idea why he had been summoned but had been told that the reason for his visit was top secret. On the journey to Moscow he had speculated that he might be designated a military attaché and sent to China. Within hours he was to learn that he was to be part of a revolution in the Soviet submarine fleet.

When he arrived at the office of the head of personnel he was ordered to sign a document stating that he would divulge nothing he was about to be told or that he would learn in the course of his work. He was then grabbed physically by the hand and taken to the navy department responsible for the testing of nuclear torpedoes and missiles. There he was informed that he was to be the second in command of the first Soviet atomic submarine which was, at that moment, being built. The commander had not been appointed, but it would be Zviltsov's job to assemble a crew and teach them how to operate the nuclear reactor that had been designed and constructed to power the submarine. Zviltsov remembered being untroubled by the responsibility of recruiting the right kind of men: 'I looked first for young and intelligent officers, whose brain curves hadn't yet straightened out and who hadn't forgotten how much you get when you multiply two by two,' he said. He was bothered, however, at the thought that, although he would be responsible for teaching the team about a nuclear reactor, he knew nothing about nuclear physics. But Zviltsov was an enterprising man even then. To master the principles of the subject so that he could lead his men, he acquired and began to read avidly *Elements of Nuclear Reactor Theory* by two Americans, S. Glasstone and M. C. Edlund, which had only recently been translated into Russian.

Less than three weeks after his meeting in the Kremlin, Zviltsov and his young team arrived in the town of Obninskoie, where the Soviet Union had located the world's first civil nuclear reactor. When he was first introduced to Dimitri Blokhintsev, head of the Institute of Physics and Energy, and Nikolai Nikolaiev, the man responsible for the nuclear reactor, Zviltsov told them that he wanted his men to observe and then man the reactor in the shortest possible period of time. He said he had been ordered to master the complexities of the reactor in two to three months and the only way to do that was for his men to take over as quickly as possible. He remembered that during this training period Alexandrov personally checked the competence and knowledge of every officer on his team. 'He would ask for answers to twenty very complicated questions,' Zviltsov said, 'and if the person failed to answer even one question he wasn't allowed to carry on.'

Stalin had been dead for more than a year but secrecy was still a way of life and the Soviet authorities tried to keep the townspeople of Obninskoie in ignorance of the fact that naval personnel were training there. However, although Zviltsov's team was only ever referred to as a group of technical and engineering workers, there were rumours that naval people were stationed at the nuclear institute. Zviltsov recalled: 'They shouldn't have known but there were rumours about who we were and why we were there. However, when the KGB got hold of those who were interested, they soon lost their desire to

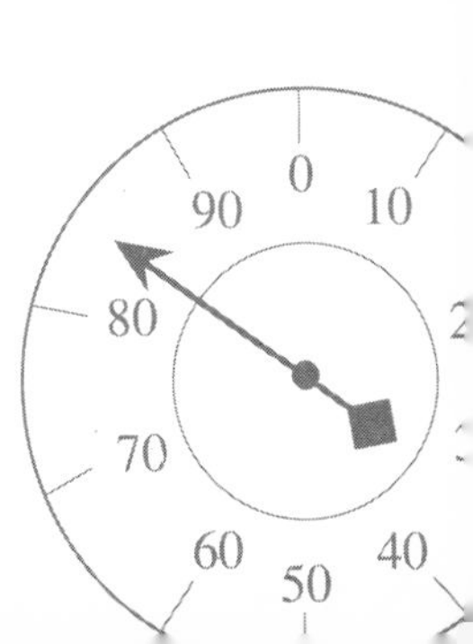

find out. The KGB also watched us and protected us from indicating accidentally that we were in fact naval personnel. We had to wear civilian clothes at all times and if, for example, someone put on trousers from a naval uniform the watching KGB officer would immediately order a change of clothes.'

Peregoudov came to meet the officers in the team and Zviltsov remembered him saying that his greatest difficulty was that he had been charged with making a boat that would perform better than the American nuclear submarines, but Soviet bureaucracy and inferior technical ability made his task nearly impossible.

Zviltsov still had no clear idea what his submarine would look like or who was going to command it – and he was already nearly three months into the project. Peregoudov had given him some clues to the boat. He said it was being designed to have two nuclear reactors and that pressurizers would maintain a level of 200 atmospheres to prevent the water boiling as the reactor heated it to a temperature of 300°C. The quality of metal in the pipework which conducted the pressurized water was very high so that the pipes would not burst while the submarine was under way and endanger the lives of the crew. Zviltsov listened carefully to Peregoudov as he went on to explain that the K-3 was designed to be at least 5 knots faster than the American *Nautilus*'s 20-plus knots. Eventually the K-3 would achieve a top speed of 30 knots by generating about 30000 horsepower, compared to the 15000 horsepower and 23 knots of *Nautilus*' and would displace 5000 tons when submerged, 1000 more than the *Nautilus*. She would have eight standard 21 inch (53 cm) torpedo tubes forward and carry twenty-four torpedoes. There would be no stern tubes and no deck guns, but the K-3 would be able to dive to 1000 ft (305 m), 300 ft (90 m) deeper than the *Nautilus*. After conversations like these, Zviltsov desperately wanted to get to grips with the submarine itself.

So did Leonid Ossipenko who was appointed commander of the K-3 as Zviltsov and his men came to the end of their training at Obninskoie. Ossipenko remembered meeting Zviltsov in the Department of Naval Shipbuilding in Moscow. 'In the corridor I saw a young man dressed in civilian clothes but who definitely looked and carried himself like a well-trained naval officer. But he had these sticking out ears, bat ears, which were emphasized even more by his closely cropped head. I asked what sort of boat I was going to command and he said, "I can't tell you now. Tomorrow I'll take you to where the crew is." I was very confused but I was in the military. I had no right to ask questions. I knew I would be told everything in due course.'

When Ossipenko eventually arrived in Obninskoie, he learnt that the boat was a submarine, but he only guessed at what kind of submarine when he discovered his crew in the final stages of learning to control a nuclear reactor. He had studied

The hand-picked crew of *K-3* on their return from the North Pole, seemingly unaffected by continually breathing radioactive air for months on end.

nuclear anti-submarine warfare and was a specialist in defence against nuclear and chemical attacks while engaged in naval warfare. He had also taken courses in mechanics and nuclear warfare and guessed that these qualifications had led to his appointment. Like Zviltsov, he also acquired a copy of *Elements of Nuclear Reactor Theory* so that he would be able to understand the basic principles of his revolutionary craft.

Ossipenko was worried about the commands he heard the civilians in charge of the reactor using on his frequent visits. He found them far too long and not suitable for a submarine in warfare. He invented a new language for his reactor crew with sharper commands that employed different and shorter words.

Although Peregoudov had finished his work by November 1954, the K-3 was not immediately translated into a steel ship. Towards the end of their training Zviltsov and Boris Akalov, one of his fellow officers, were taken to Leningrad to see complete sections of the entire length of the submarine all built in wood on a 1:1 scale. The five separate sections of the boat were in secret locations all over Leningrad so that no one could stumble upon the complete submarine model and divulge the secret. One of the sections was closely guarded in a basement near the Astoria Hotel. Zviltsov remembered it clearly: 'Today in St Petersburg it's something they'll tell you never happened. But they built it in wood to see how it could be fitted out, how the reactors could be installed and how to arrange the equipment in the best possible way.'

Zviltsov and Akalov discovered they were the naval representatives on the Model Commission and had the responsibility of making sure that everything fitted. They were told that if something did not fit, they had to make it fit. Ossipenko later recorded: 'For the designers, the first priority was the weapons systems, the armaments, the rockets, the torpedoes; then came the nuclear plant; then came technical and other problems. Habitability, convenience and comfort were their last priority.'

The appointment of Zviltsov and Akalov to the Model Commission was Peregoudov's idea. He wanted the submariners to have the best conditions possible. He wanted life on board the submarine to be amenable so that the crew would be able to make the most of the boat and its revolutionary design. But he understood that such a move would not please everyone. He confided to Zviltsov that the concept of individual comfort, especially for military personnel, went against Communist philosophy and the beliefs of the political leaders in charge of the project. 'Inspect all the compartments on the K-3 mock-ups,' he told Zviltsov, 'all the living areas and washrooms and come back and tell us how we can arrange them in a better way. Go and see how the cabins of sleeping compartments on trains, and passenger cabins on liners and aeroplanes, are equipped. Bring back anything that we can put into your submarine that will make you more comfortable. Don't hesitate to ring me if there is something at fault on board.' He gave Zviltsov his direct telephone number, adding, 'For you I am there at any time.'

So Zviltsov and Akalov began to study the layout of the craft in which they were likely to be submerged for months at a time. As they travelled around Leningrad they found ropes of different colours laid throughout the various model sections to represent the intricate cabling of the K-3. Their best weapons were handsaws. In section after section they literally sawed pieces of wood representing equipment or furniture off the models and rearranged them where they wanted them to be. 'Why have you installed this piece of apparatus in front and that other piece behind?' they would ask designers. 'In a submarine they should be side by side.'

They soon realized that there had been no submariners on the design committee. 'The construction designers had put things where there was a place available without thinking about what they might be for or the circumstances in which they would be used,' Zviltsov recalled. In the control room section in one of the Leningrad hide-outs the captain's and observer's positions were facing aft towards the stern rather than towards the bow, so that when the submarine was under way the captain would be forced to reverse all his directional commands in his orders to the helmsman.

Zviltsov got that position turned around, and then found that the section containing the officers' wardroom had been designed so that only seven of the thirteen officers off watch at any one time could be accommodated around the table. 'We enlarged the wardroom by pulling down one of the partition walls and extending the space available,' Zviltsov recalled. 'Boris and I cared because we were going to use it later, but others – we called them heels or old farts – disliked what we were doing. They had been used to diesel-electric submarines in wartime when conditions were very tough but when, at least, they could surface quite often. But they were too out of touch with the problems of everyday life in nuclear submarines where better facilities were needed to cope with living submerged for long periods of time.' By the time Zviltsov and his colleague had finished, the K-3 wardroom was fitted out with good and expensive furniture and boasted a beautiful limewood table.

Zviltsov also fought to get a refrigerator installed on board the submarine rather than the cold-store cupboards he found there. Traditional submariners were resistant once again but the refrigerator eventually appeared. When the crew learned that the K-3 might not surface for weeks on end during a mission they also put forward suggestions. By the time they had finished not only had the living conditions changed considerably but the refrigerator was well stocked with caviare and vodka. The crew liked caviare; they also liked vodka. But during the early voyages of the K-3 the vodka was to be used extensively for medicinal purposes to counter the effects of radioactivity.

Ossipenko had known the technical parameters of *Nautilus* in 1955. 'At that time,' he remembered, 'newspapers were writing about the adventurous approach taken by American designers in constructing *Nautilus* and *Skate*, the first nuclear submarines. The American design, though, had been criticized for having too little buoyancy. It was said that the submarines wouldn't sail well. But we had our own feelings. The Americans are not such idiots as to build bad boats but we didn't have to imagine what they looked like. We simply read the press. Everything was clear about them.'

What Ossipenko had not appreciated was the effect the completed K-3 would have on him when he first saw her waiting to be launched on the stocks at Severodvinsk in May 1957. Although he had studied the sections in Leningrad he had never come near to anticipating the lines of the boat he saw in front of him. 'I was very impressed by its style and immediately realized what a powerful and beautiful creature it was,' he recalled. 'The very big stern and the teardrop shape added to its solidarity and gave the impression that it was a really powerful boat. It measured 105 metres long and had two big screw propellers. I knew it was high time to leave the simulator and move on to the boat and get to grips with it.'

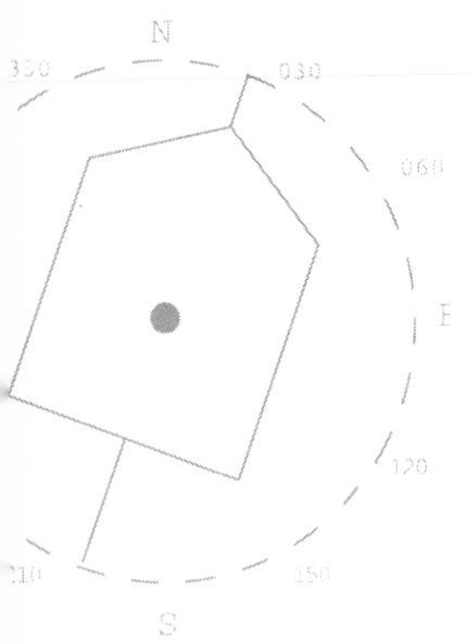

When the K-3 was finally completed in August 1958 it took Ossipenko less sea-time to get to know her than it had taken to familiarize himself with any of his previous submarines. On the very first trial he stayed submerged longer than he had planned and the design staff and engineers on shore began to worry. They were about to begin a rescue operation, thinking the vessel had sunk, when the K-3 resurfaced.

Zviltsov has different memories. He recalled that some things were perfect: 'When, in the tests, the reactor drove the submarine to standard speed, everyone on the bridge was shaken by the quietness. For the first time in all my duty on submarines, I heard the sound of the waves near the bow end. On conventional submarines, the sound of the exhaust from the diesel engines covers everything else. But here there was no rattling and no vibration.' But other things were troubling: 'The biggest problem facing us was the full reliability and security of the nuclear plant and of the turbines. In the event there were no major accidents, no casualties, but there were many things which went wrong. On board for the trials were many representatives of the shipyard which had built the K-3 and designers who could see with their own eyes what needed to be done to protect the crew with a radiation shield. The officers who had undergone training in Obninskoie knew they had already received doses of radioactivity. They no longer worried.'

In the early days of nuclear projects there was little accurate information about minimum dosage or the long-term effects of radiation. But Zviltsov knew that the problem of radioactivity on board the K-3 would not come from direct radiation but through the water supply and the release of radioactivity from the water into the air. The weak point was where the heat-bearing contaminated water circulating through the reactor system came into contact with the pipes that conducted the heated pure water in the secondary system around the boat and into and out of the living quarters of the crew. If there were any leaks from the first system into the second, the water in the second one would also be contaminated.

'It turned out that the air inside the submarine in the early days was full of radioactive sprays and gases,' Zviltsov recalled. 'It must have been dangerous but no one felt ill as a result of that. The funny thing was that after a term of service in the nuclear submarine, the men were only worried about the sex of their children. Before serving aboard nuclear submarines they had had daughters; afterwards they had sons.' Then Zviltsov became serious. 'They knew they were at risk but Ossipenko went on with the trials. They all knew they had to go on because without their knowledge the task couldn't be completed. Ossipenko decided during the trials they would carry on sailing until the radioactivity equalled 100 permissible doses. That is 100 times the maximum permissible dose – when maximum equals safe.'

Though it may sound extraordinary today, Ossipenko decided that the air in different compartments of the submarine should be mixed up in order to reduce the effect of this radiation. The air in the compartments housing the reactor and turbines was more radioactive than in some others and he ordered that an air draught be created through the length of the craft so that any radioactivity would be divided and distributed equally. He also ordered that the hatches should be opened as soon as the submarine surfaced to allow the boat to be ventilated.

Dr Ivan Bechik, the medical officer aboard the K-3, was charged with keeping the crew healthy and free from radioactivity. When he went aboard during the trials he was informed that radioactive gases were reputedly leaking from a steam generator. That particular generator was switched off when it was identified and the boat continued on the other generators. But Dr Bechik was never worried. 'Conditions never got to the really critical level, but when the readings showed that the radioactive level was higher than normal, the submarine surfaced and the hatches were opened. The submarine was then ventilated or we had to go back to the naval base. The agreed limit of 100 permissible doses within a short period of time is not dangerous. It is dangerous to get 100 permissible doses over a longer period of time. The radioactivity didn't really affect our health and you can see now that all the members of the first crew have survived and are in good condition. I cannot paint a very black picture of how it was inside the submarine.'

Zviltsov remembered that everyone looked towards the refrigerator for protection against radiation poisoning. The crew ate cod-liver pâté and drank vodka. Both were thought to be effective but he still praises the properties of vodka: 'One hundred and fifty grams of vodka after a day's work eliminated any radiation effects and restored the metabolism,' he recalled.

But international events and Soviet pride and ambition were to override the health concerns of the sailors on the K-3. At 11.15 p.m. on 3 August 1958 USS *Nautilus*, the world's first nuclear submarine, reached the North Pole. The news of the historic voyage under the polar ice cap excited the world. But in Moscow, Premier Khrushchev had taken it badly. He had ordered the Soviet nuclear submarine commanders to catch up and pass the Americans' achievements. The progress of Ossipenko and Zviltsov and the crew of the K-3 was about to be closely scrutinized as they began to train, practise and prepare their boat to follow *Nautilus* to the North Pole.

Zviltsov succeeded Ossipenko as commander of the K-3, known by then as

ABOVE A symbolic greeting for the *K-3* and its crew at the North Pole.

RIGHT Moment of triumph. *K-3* at the North Pole three years after *Nautilus*. The first part of Khrushchev's order to catch up and pass the Americans was close to being realized.

Leninski Komsomol, in 1959. At the time he had not been especially excited to be given command. Two years later he brought his radioactive, fault-ridden craft back from the North Pole to be hailed as a hero of the Soviet Union. In the months which followed, as news of the achievement leaked out, the Americans found it hard to believe that, after the lead they had once enjoyed, the Russians were so close behind them in nuclear submarine technology. Moscow eventually announced the news on 23 October 1961. Fifteen months later, when the Soviet Union sent its second submarine to the North Pole, the Americans realized that their advantage had all but gone.

In little more than ten years submarine designers with no knowledge of nuclear physics, nuclear engineers who knew nothing about submarines, propulsion engineers kept ignorant of the project on their drawing boards, industries who supplied parts for unknown machines and sailors who drank vodka to ward off radioactivity had created and perfected a war machine that generations of men had coveted for centuries.

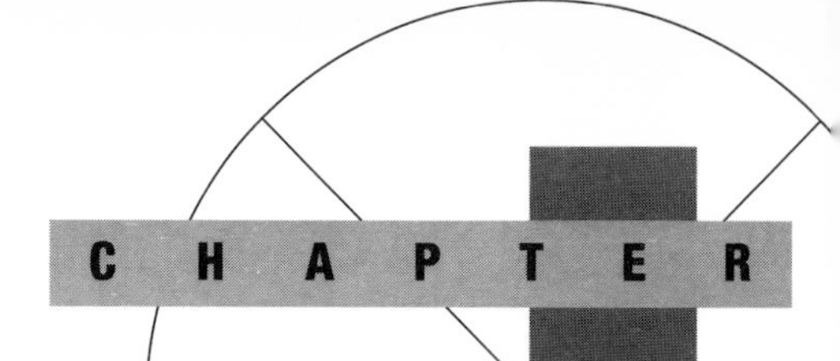

2

THE EARLY PIONEERS

Those American and Soviet designers and nuclear scientists who perfected the first genuine submarines after the Second World War are likely to have been far more familiar with the life and times of Albert Einstein, than those of David Bushnell, Robert Fulton, John P. Holland and Gustave Zédé. And yet the development of the submarine seems to have had little to do with man's idiosyncratic dreams of conquering the deep and everything to do with a general hatred of the British and their empire.

It was such a hatred that feelings of outrage and the need for revenge burned deep inside those who felt they and their countrymen had suffered; a need fuelled by folk memories of brutality in Ireland, punitive exploitation of settlers in the new colonies of America and a military and political domination that had humiliated the French for centuries. Such obsessions inspired a succession of brilliant inventors to try to perfect a secret weapon capable of destroying the battleships of the Royal Navy, with which Britain had built and maintained her empire since well before the middle of the eighteenth century.

And most of those men would have smiled wryly and knowingly if they could have seen the speed with which Britain, having contributed least to the direct development of the submarine before the beginning of the twentieth century, came to embrace this weapon of revenge she had inspired once she realized its true potential.

It is ironic, therefore, that serious thinking about a craft which could travel under the sea had really begun with an English enthusiast in the sixteenth century. In 1578 William Bourne wrote *Inventions and Devices*, a book that must have been essential reading for the generations of inventors who followed him. 'It is possible,' he claimed, 'to make a shippe or boate that may goe under the water unto the bottome, and so to come up again at your pleasure.'

Bourne described a system of horizontal screws connected to leather pads which, when unscrewed, would allow sea-water to enter chambers on the side of the submarine causing the craft to sink and which, when screwed the other

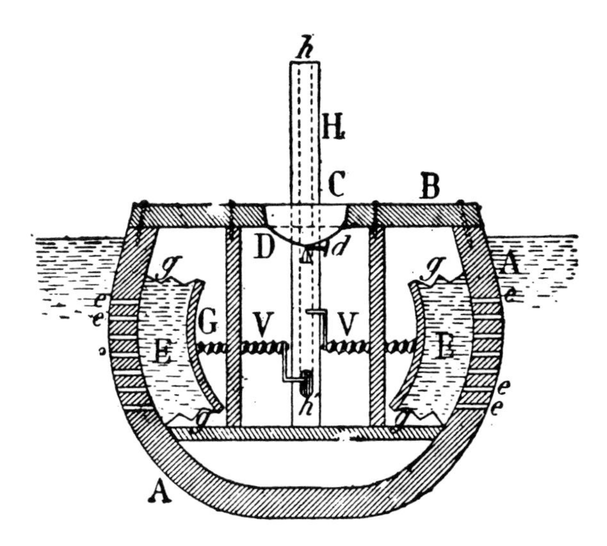

William Bourne's first principle: flood chambers, with water to submerge, expel the water to surface. 'It is possible to make a shippe or boate that may goe under the water unto the bottome, and so come up again at your pleasure.'

way, would expel the water causing it to rise again.

Nearly 200 years later David Bushnell, from the state of Maine in New England, wrote about his idea of an egg-shaped craft designed to take on board just enough water-ballast to allow it to travel fully submerged below the surface of the sea. Bushnell first described his battleship-wrecking device in 1754, little more than a year before the start of the American War of Independence between the colonists and the British. The Victorian drawings of Bushnell's craft, *Turtle*, show a man, seated, turning the horizontal screw while looking through a porthole and controlling the rudder by jamming it under his arm. Once the hatch was closed the one-man crew had sufficient air for only 30 minutes of effort. With one of his feet he pushed a lever which opened a valve and allowed water to flow into the vehicle. Once a condition close to weightlessness in the water had been achieved the *Turtle* could be screwed forwards or backwards, or even up or down. A weapon of great potential had been delivered to the colonists. It was only once sent into action against the British fleet during the American War of Independence when Sergeant Ezra Lee used it to attack the battleship HMS *Eagle* anchored off Governor's Island, New York. The *Eagle* survived the attack, but the British understood the threat presented by the small craft. Without ever realizing how limited it was, they were suddenly aware that some new enemy device had the capacity to approach and attack their ships while completely submerged. Soon afterwards Bushnell went to France to try to interest the French in the *Turtle*. His accomplishments had received a great deal of public attention, but although the authorities believed that a development of undersea warfare might ensure victory in wars with England yet to come his ideas were not accepted.

In October 1797, the French Republic, 'one and indivisible', had no enemy but Britain, whose policy was to blockade French coasts and ports with her superior fleet. The French government, faced with such constant humiliation, was forever searching for some means by which they might be delivered from the domination of the British.

One man who might have helped solve their problem was Robert Fulton from Little Britain Township, Pennsylvania, who had also designed a submarine. Fulton was interested in political ideas and, as a citizen of the United States of America, held the fundamental belief that world peace could be achieved only if commerce was allowed to move freely around the world, with no maritime force able to deny merchant ships a port of call.

By 1797 he was convinced that to achieve such peace it would be necessary to get rid of all navies. In particular he believed that the power of the British navy should be diminished. The French, despite the problems presented by the British blockade of their ports, were not inclined to encourage the American, who had appealed to one minister: 'The destruction of the English navy would ensure the independence of the seas and also of France which alone, and without rivals, holds the balance of power in Europe.'

Fulton finally persuaded the French to let him produce a submarine for them. He had long since decided to call it *Nautilus*. The vessel was 21 ft (6.5 m) long with a 7 ft (2 m) beam and built in the shape of an ellipse with a copper shell covering an inner frame. In 1800 it was successfully tested in 25 ft (7.5 m) of water at Le Havre on the west coast of France. Fulton adapted the foot-operated valve that Bushnell had used to control the amount of water-ballast taken on board and also incorporated Bourne's idea of flooding special reservoirs to submerge the craft and pumping them out later to allow it to resurface.

In the late summer of 1801 *Nautilus* was demonstrated to watching crowds on a stretch of the Seine in Paris, above the Hôtel des Invalides. Fulton and a sailor carrying a candle submerged for 25 minutes and travelled a considerable distance under water at a rate of about 1½ miles (2.5 km) an hour before resurfacing. They also travelled up and down several times. For both demonstrations his craft was powered only by the muscles of the two men. Their strength alone turned a screw at the stern.

A model of *Turtle*, the first midget submarine, which was capable of approaching and attacking enemy ships while remaining mostly submerged.

His second *Nautilus*, built soon afterwards, used the same principle of power but incorporated a massive bar of metal which acted as a keel and counterweight. In 1802 Fulton agreed to run tests for the French to see what might happen if the second *Nautilus* came up against ships of the British navy. The craft was put through its paces at Brest where the French *Naval Chronicle* reported Fulton as having: '... not only remained a full hour under water with three of his companions, but held his boat parallel to the horizon at any given depth. He proved the compass points as correctly under water as on the surface, and while under water the boat made way at the rate of half a league per hour by means contrived for that purpose. Mr Fulton has already added to his boat a machine by means of which he blew up a large boat in the port of Brest: and if by future experiments the same effects could be produced on frigates or ships of the line, what will become of maritime wars, and where will sailors be found to man ships of war, when it is a physical certainty that they may every moment be blown up into the air by means of a Diving Boat against which no human foresight can guard them?'

It is difficult now to imagine the fascination with submarine boats at the beginning of the century. Fulton's *Nautilus*, depicted under sail on the surface, was a collector's item.

In practice Fulton, like Bushnell before him, suffered from the distaste with which French naval traditionalists viewed the prospect of submarine warfare. The maritime prefect of Brest issued orders that he was to cease and desist from his idea of attacking English ships, because: '... such warfare carries the objection that those who undertake it and those against whom it is made will all be lost. This cannot be called a gallant death.' The French Minister of War, Admiral Plevelle le Pelly, wrote that it was impossible to: '... serve a commission for belligerency to men who employ such a method of destroying the fleet of the enemy'. Even the Emperor, Napoleon Bonaparte, branded Fulton a fraud and a man without morals.

The British, however, were taking a much more pragmatic view. In a secret circular sent from the Admiralty to five of its flag officers it reported: 'Mr Fulton, an American resident at Paris, has constructed a vessel in which he has gone down to the bottom of the water and has remained thereunder for the space of seven hours at one time – that he has navigated the said vessel, under water, at the rate of two miles and a half per hour; that the said submarine-vessel is uncommonly manageable, and that the whole plan to be effected by means thereof, may be easily executed, and without much risk. That the ships and vessels in the port of London are liable to be destroyed with ease, and that the channel of the River Thames may be ruined: and that it has been proved that only twenty-five pounds weight of gunpowder was sufficient to have dashed a vessel to pieces off Brest, tho' externally applied.'

But Britain never encouraged Fulton directly, despite the fact that he went there to offer the Admiralty his experience – his political idealism, it seems, conveniently forgotten. He eventually returned to America where, unsung, he turned once more to his drawing board and came up with a design for the paddle-steamer – an icon of peaceful and gracious travel for which, rather than a silent, violent deliverer of death, he is today mostly remembered.

Sixty years later, during the American Civil War, the Southern states, home of the paddle-steamer, took the lead in the development of submarines propelled by steam-engines on board or by compressed air held in large reservoirs. Submarines increased in size and in horsepower, until by 1865 *Spuyten Duivel*, a semi-submersible, was 82 ft (25 m) in length, 20 ft (6 m) in the beam, displaced 270 tons and was protected by 2 inch (5 cm) armour plating. It could carry,

OVERLEAF Sergeant Ezra Lee, piloting *Turtle*, fails in his attempt to blow up HMS *Eagle* during the American War of Independence. Painting by D. A. Ropkins.

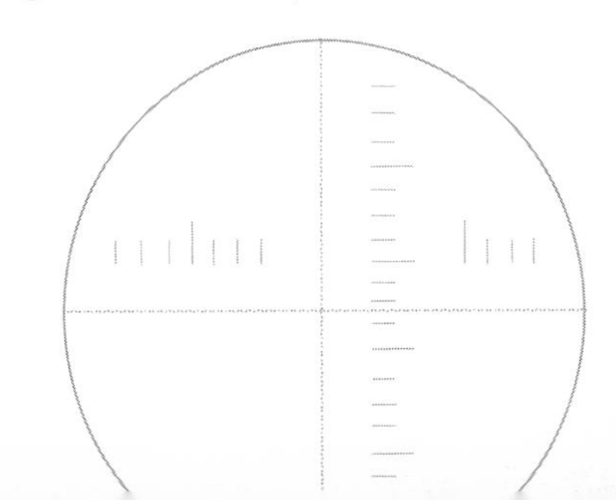

but not fire, a torpedo; although the first Whitehead torpedo had been invented and fitted to surface craft, it was to be some time before it was launched from special tubes fitted on board a submarine.

The Confederacy also introduced steam-powered vessels trimmed to ride low in the water and designed to travel at top speed to ram and hole enemy ships below the water-line. One of these, the *Manassas*, was whale-backed with a huge stack like some strange dorsal fin. She was 128 ft (39 m) in length and her first 20 ft (6 m) of bow was solid iron. She displaced 387 tons and a cannon anchored just behind her solid nose was capable of firing balls weighing 68 lb (31 kg). This craft, monstrously large for its day, was said to have been responsible for the North's warships finally abandoning the Mississippi river to the Confederate forces.

It was France, however, who made the greatest contribution to successful submarine design between 1860 and 1880. Development there was particularly intense, because designers and engineers received government backing as a result of the thinking of the Jeune Ecole, a new school of French naval strategists who advocated using small torpedo boats – a definition that came to include submarines – to keep their ports open and, if necessary, defend France against British naval supremacy.

Oliver Riou, an engineer, first suggested, in 1861, that electricity from batteries should be used to power submarines, but it was another ten years before his ideas could be considered seriously. Early electric storage batteries were heavy, inefficient and worryingly prone to leak poisonous fumes.

By the late 1870s the French were confident that no other nation could claim greater knowledge of submarine technology. But 3000 miles (4825 km) away, across the Atlantic in New Jersey, a small, short-sighted Irish teacher with a walrus moustache, who had stepped off a boat early in the decade to begin a new life in America, nursed his own, quite definite, ideas about how submarines should be developed.

THE RACE TO BE FIRST

John Philip Holland watched the team of eight pairs of horses, attached to the heavily loaded wagon, back up to the edge of the river. Sitting securely on the wagon was his very first submarine – a big, black mass of iron with a turret – which was to become known as *Holland* I. It was 14½ ft (4.5 m) long, 3 ft (90 cm) wide and only 3 ft (90 cm) high.

The scene had drawn a large crowd which watched intently as the wagon-master untied the ropes that held the submarine in place, and allowed the 2½ ton craft to slide into the water. The spectators saw the vessel submerge, then surface, then stabilize itself – and then slowly sink from sight. Some time later Holland watched his dream dragged out of the shallow water to rest in the thick mud of the river bank and began to investigate exactly what had gone wrong. He knew he did not have long to find the answer, because his strange and secretive financial backers, with their own far-fetched but intensely serious plans, would also have been watching events on the Passaic river in Paterson, New Jersey, that morning in 1876.

In November 1873, when Holland had arrived as an Irish immigrant teacher in Paterson he could have had no idea that he was fated to become the most important submarine designer in the United States of America. Nor could he have imagined he was about to begin a race with a series of French designers for the right to be called the father of the modern submarine.

He was born in 1841, the son of a British coastguard service officer in the tiny town of Liscannor, County Clare, at a time of great misery and hardship in the west of Ireland. The great potato famine had caused starvation among hundreds of thousands of his countrymen and before he was ten years old, one of his three brothers and two of his uncles had died of cholera. Peasants who had worked for absentee landlords lived in abject poverty, and some of those without jobs were evicted from family homes which were then stripped of their roofs so that they could not be reoccupied. Such events led to a mass migration to the United States by more than a million people in the years between 1847

and 1854. With them the Irish carried a resentment of the English that has never been forgotten.

John Holland was never to escape this legacy. When he was seventeen he trained as a teacher with the Order of the Irish Christian Brothers, a Roman Catholic teaching order which was well known for its sympathetic attitude towards the dissolution of the union with England. His older brother, Alfred, was involved in printing a semi-revolutionary weekly in Dublin and his younger brother, Michael, was a declared and active separationist. Michael Holland fled Ireland for the United States and was followed first by his parents in the spring of 1872 and then by John, who had withdrawn from the Order of Christian Brothers, in 1873.

When John Holland arrived in Boston he had already made sketches and calculations for a submarine. Years later he recalled the feelings which motivated him at that time. 'I knew that in a country where coal and iron and mechanical skill were as plenty as they were in England, the development of large armour-plated ships must come first. Therefore I must get to a place where mechanics in shipbuilding were less advanced, and the available material for big iron-clad vessels scarcer. Then, too, I was an Irishman. I had never taken part in any political agitation, but my sympathies were with my own country, and I had no mind to do anything that would make John Bull any stronger and more domineering than we had already found him.'

While working as a lay teacher with the Christian Brothers in Paterson, Holland found time to develop his ideas for a submarine and they were considered by officers at the naval torpedo station in Newport, Rhode Island. The navy told him that no one would go down in such a craft as he had designed and that Holland would do well to drop the whole matter.

Holland's designs might well have stayed no more than lines and calculations if the Irish question had not forced its way back on to the political agenda in Ireland and Britain and into the Irish communities in which Holland was now moving. In 1871 Gladstone, the British Prime Minister, had exiled fourteen Irish revolutionaries who had been held in British prisons. By the time they arrived in New York they were heroes. Two of them, John Devoy and Jeremiah O'Donovan Rossa, became central figures in the Fenian movement that was

John P. Holland who had his own, quite definite ideas about how submarines should be built and who soon found himself in a race with the French.

already established in the United States and provided an energy that led to the Irish question becoming a topic throughout America.

Shortly after Holland's ideas had been rejected by the US navy his brother Michael introduced him to Rossa and then to Devoy. The latter wrote of him: 'He was well informed of Irish affairs and was anti-English and with clear and definite ideas of the proper method of fighting England. He was cool, good-tempered and talked to us as a schoolmaster would to his children.'

It is not hard to imagine, in the light of the lengths the present IRA is prepared to go to attack the British Establishment, how the more revolutionary of the Fenian leaders would have been excited at the idea of owning a weapon such as a submarine, which could get near to one of the huge British warships without being seen and unleash a charge which could take her to the bottom of the sea. It would be relatively cheap and was exactly the kind of weapon they needed. Moreover, in 1875 Rossa had set up a 'skirmishing' fund to which Irish men and women of all occupations in the United States contributed hard-earned money, so great was the hatred of England among the émigré community.

The following year Holland heard that the Fenian leadership were prepared to give him the financial support he needed to make his first submarine. The Iron Works on Albany Street, New York City, was given the contract. To keep the project secret, code names were given to Rossa and James Breslin, another Irish hero who was to be Holland's day-by-day contact with the Fenian organization. The boat, scheduled to cost $4000, was built in Paterson and an internal combustion engine, patented by George Brayton only four years earlier, was installed in the craft.

When John Holland's first submarine sank the day it was launched the crowds must have believed they had seen the last of what Holland called his 'wrecking boat'. But the Irishman was not about to give in at the first attempt – and he had his financial backers to convince.

Two weeks after the disastrous first launch Holland slipped through the turret and folded himself into a space 3 ft (90 cm) wide, less than 4 ft (1.2 m) long, and a little over 2 ft (60 cm) high. His eyeline was level with a small window in the closed turret. On a command, he took on steam at pressure from a nearby launch and moved gently away upstream. A few minutes later he flooded two tanks, adjusted the diving planes and slid under the surface of the river. When the craft emerged a short distance away, Holland popped out of its turret smiling broadly. His boat worked. It could run on the surface, submerge and resurface.

That short exhibition indicated to him that a successful submarine needed a constant reserve of positive buoyancy (two of the empty tanks on *Holland I* were watertight) and a low and fixed centre of gravity to ensure stability. He also learned that the hydroplanes needed to be moved from the centre of the craft back to the stern adjacent to the rudder to make the act of submerging easier. The watching Fenian officials realized they now had a potent weapon on their hands but they did not all agree on when it should be used. Since they had no immediate use for the craft they decided it should be scuttled in the Passaic river until it was needed to be used against the British.

However, they wanted Holland to continue his research and early in 1879 the inventor travelled to the Delameter ironworks on West 13th Street, New York City, to talk to the owners, Cornelius H. Delameter and George H. Robinson. He wished to discuss the building of the successor to *Holland I*, which was to have the capabilities of a fully operational war machine. Among other things, Holland wanted to be assured that the two men could keep the project secret. He also explained that they would have to trust him; he would not even be able to tell them who was putting up the money to pay for the new craft.

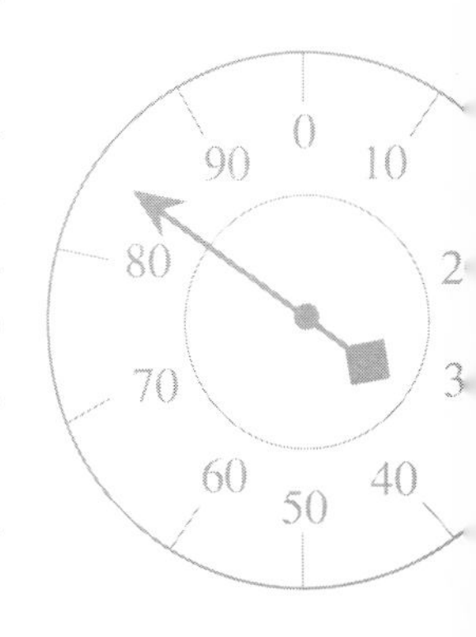

When he finally unrolled his drawings and specifications and Delameter and Robinson saw that the cigar-shaped boat was designed to rest in the water three-quarters submerged, they must have wanted to ask many more questions than Holland could have answered. When he left the ironworks its owners had agreed a price of $20 000 but they must have been unconvinced that such a craft could be launched without sinking. It was a feeling that spread among workers at the foundry and led to disruptions and delays after construction work began on 3 May 1879.

Emissaries from Germany, Italy, Sweden and Russia came to see the building of the boat at the Delameter works. There were also constant visits from Holland's Fenian backers and, on one occasion, two Turkish envoys who, Holland noted, acted without the caution and secretive manner other visitors showed. 'But, very clearly to me, they had no idea of the importance of what was expected from the machine, or, much more likely, they had been persuaded by their acquaintances of English connection that the project would never amount to anything because it did not originate in England,' he was to write.

Holland would have been much more worried if he had known about the attention an English visitor to the Delameter works was paying to his latest invention.

The British Consulate in New York heard that the Fenians were financing the project and in March 1880 began to take an interest in Holland's new construction. Captain William Arthur, the naval attaché, was initially sceptical of the rumours that a submarine was being constructed but between March and May that year he visited the Delameter works and became convinced that this was indeed the case. A private detective was even hired to keep track of the boat's progress at the

ABOVE The *Fenian Ram*, John P. Holland's first successful submarine, named after his Fenian backers, on display in Patterson, New Jersey.

TOP Interior of the *Fenian Ram* showing the drive mechanism.

ironworks and he established a personal relationship with Cornelius Delameter – who allowed Captain Arthur secretly to copy Holland's plans. Even the US Customs Service was persuaded to keep watch on the submarine on behalf of the British: 'The American government will do anything to carry out the wishes of Her Majesty's Government with regard to this and any other such plans,' the British Vice-Consul in New York telegraphed to London on 3 September 1881. The message confirmed the greater interest the Foreign Office was taking in what had become dubbed 'The Fenian Ram'.

The Foreign Office was in the middle of a sustained major intelligence operation against the Fenians and had targeted several individual societies for observation. A variety of terrorist attacks in England had convinced the British government that the Fenians were a threat which had to be taken seriously. For the diplomats in New York, John Holland's links with the skirmishing fund made the project something that needed to be watched very carefully. In Britain the Admiralty was more interested in the submarine's technical shortcomings than in her political significance and did not see the submarine as much of a threat. The Director of Naval Construction wrote in June 1880: 'There seems no reason to anticipate that this boat can ever be a real danger to British ships ... We should not recommend the spending of any money in order to obtain information.'

By the time the boat was ready to be launched two years later, Holland's Fenian backers were arguing openly among themselves about what role their movement should be adopting against England. Devoy had come to believe that honourable, open warfare against the English was futile. Rossa agreed with him absolutely and the whole nature of the skirmishing fund was under examination. Devoy was accused of failing to respect the original intent of the donors, of funding open and rebellious acts against British tyranny. In defending his colleague, Rossa, exasperated, wrote to one of the fund's trustees: 'You want ' "honourable warfare". Well, wait until England will let you have it and you'll wait till you'll lie down and die.'

It has been estimated that by the time the new submarine was ready to be launched, almost two-thirds of the entire fund had been spent on Holland's projects alone.

The *Fenian Ram* was first seen publicly at the Morris and Cummings Dredging Company's dock in Jersey City, across the Hudson river from the Delameter works. She was 31 ft (9.5 m) overall, with a 6 ft (1.8 m) beam and measured more than 7 ft (2 m) high. Probably the most impressive feature of the craft was her lines. Holland had set out to replicate the configuration of the porpoise in an attempt to achieve hydrodynamic efficiency.

Holland had calculated the ramming ability of the 11⁄16 inch (2 cm) thick iron hull, driven by an improved Brayton petroleum engine, as nearly 19 tons. In her trials she accidentally rammed the Morris and Cummings pier and not only split a 12 inch (30 cm) pile, but lifted a horizontal tie bearing a load of 4 ft (1.2 m) of stone ballast. Holland claimed that the only thing damaged was his engineer's respect for good English.

Within a week the *Fenian Ram* had successfully completed all her tests under water. She had behaved like a porpoise in diving and surfacing at an angle. The engine worked well at depths of 40 ft (12 m) or more, and the two-man crew felt no ill-effects since the air for the engine was released continually from a storage cylinder. Holland considered her a great success. 'There is scarcely anything required of a good submarine boat that this one did not do well enough, or fairly well,' he wrote later.

But there were disputes over money and payments within the ranks of the Fenian brotherhood, and between the Fenians and Holland. To pre-empt a court decision which might have gone against them, the Fenians one night stole the craft from her moorings on the Hudson river and took her to New Haven, Connecticut. Holland was disgusted and vowed to let her rot on their hands. He was never to work with the Fenians again.

The period which followed must have been one of great frustration for Holland. His next submarine design came to nothing, the victim of a failed business venture, just when advances in submarine design were emerging from an entirely different tradition, in France.

France was the only European nation intent on developing submarines. There had been some interesting, individual efforts in Britain and Russia, but none of them rivalled the designs that Claude Goubet and other French designers produced in the last twenty years of the nineteenth century.

Claude Goubet's first boat was more than 16 ft (5 m) long and nearly 3½ ft (1 m) wide. It weighed 2 tons and its power unit was based on banks of storage batteries. It was launched at about the same time as the *Fenian Ram* but, unlike Holland's submarine, it led to further developments. Once Holland stopped production after the *Fenian Ram* no one else took on the mantle of leader of submarine design in the United States and the development of the internal combustion engine as a power source for submarines came to an end. In Europe, on the other hand, Goubet among others was demanding the latest technology from electric battery manufacturers to allow them to design craft for longer underwater journeys.

Goubet's electric motor worked well enough, but the two crew sitting back-to-back in the centre of the craft had difficulty controlling the depth at which

Claude Goubet's first submarine which ran under water on banks of electric batteries but which had difficulties in maintaining a constant depth.

the submarine ran under water and its stability from fore to aft. The problem was one of balance. The angle of dive was regulated by the movement of the mass of water inside the boat. The boat submerged to a pre-set depth by taking water into central ballast tanks. Theoretically, it would stay horizontal under water because water would be pumped fore or aft as required to balance the trim of the boat. Sadly, an automatic device controlled by a pendulum failed to operate the pump accurately enough and the boat, cast as a single shell of bronze, was not a success.

By the summer of 1889 Goubet had produced the *Goubet II* which, again with a crew of two, rivalled the *Fenian Ram* directly in its engineering design and its capability both under and above water. Those eight years had been profitable for Goubet. The electrical equipment in *Goubet II* operated much of the apparatus incorporated in the boat. Goubet even linked the power to a 'headlight' on the bow. He had also solved the problem of longitudinal stability which had plagued his first craft. However, he still had to find a solution to the problem of running at a constant depth under water.

During these years Holland would have heard that despite the boat's performance, the French Government declined to make any investment in *Goubet II* and that another French inventor, Dupuy de Lome, had also designed

a submarine but had died before he could see it completed. Some time after his death, the plans for the craft de Lome had called *Gymnote* were seen by the famous naval engineer and shipbuilder Gustave Zédé, who used contacts inside the French navy to have the project turned over to a firm working on the French Mediterranean coast. When *Gymnote* was launched in September 1888 the French believed they had the best submarine the world had yet seen.

The craft was 56 ft (17 m) long and her slender hull, designed as a cylinder with conical ends, displaced 30 tons. She was thoroughly braced to withstand pressure, and the current to drive her massive electric motor, which delivered 55 horsepower, was drawn from a bank of 564 storage batteries. It made Goubet's power unit of eight years before look puny. At her launch she could only make 8 knots on the surface compared to the 9 Holland had achieved with the *Fenian Ram*. However, she had two horizontal rudders near her stern and these gave her the ability to maintain a constant depth and steer a straight course under

Gymnote, believed by the French to be the best submarine the world had then seen. Holland was scornful and suspected industrial sabotage of his own designs.

water – two characteristics that earned her the accolade of being the first modern submarine. France had no hesitation in commissioning *Gymnote* into her fleet – the first submarine ever accepted by a major naval power.

Holland was scornful when he heard of her success. He pointed out that the submarine could not proceed far from her base of operation because she had no means of re-charging her batteries. He also suspected espionage: 'About this time the United States Navy Department was mildly interested in the performance of submarines in France, where they had attained some slight degree of success. The designs of these boats, I am sure, were based on certain fundamental points of my *Fenian Ram* design. As I have said previously, there were a number of foreign officers present at Delameter's yard while the boat was in the course of construction, and it is hardly to be expected that they failed to take notes. However, the knowledge they secured did them very little good, because, while they secured a lot of valuable data, their inexperience caused them to disregard the most vital points, with the result that their boats never attained any degree of success. However I do not wish to convey the impression that the United States Navy Department was at this time considering building submarines as a result of the French experiments; far from it. Had it not been informed of the success of my *Fenian Ram*, which was far more interesting and wonderful than anything the French had done, and still remained unconvinced? I was totally sick and disgusted with its action, and was seriously tempted to abandon all further attempts to convince and awake it from its lethargy.'

It is undoubtedly true that Holland felt himself to be in competition with the French designers. He was by now a bitter man. The US Navy Department was not interested in his submarines. His backers had gone. He had been forced to take a job as a draughtsman to keep his family after the *Fenian Ram* affair and his latest submarine design had come to nothing. He felt alone, abandoned and totally frustrated.

Paradoxically enough, in 1888, soon after *Gymnote*'s launch, Holland got the break he had been waiting for. Possibly by coincidence, but more probably as a result of the beginning of a French submarine service, the US Navy Department invited entries for a competition to produce a submarine torpedo-boat capable of moving at 8 knots under water and 15 on the surface, which could run at maximum speed under water for two hours and which would have the capacity to fire torpedoes. When Holland won the competition it must have seemed that the lean years were over. However, it was another five years before the US Government found the $200 000 needed for its construction. And at the beginning of July 1893, as John Holland arrived in Washington to discuss his prize-winning designs, news began to filter through that the previous month the French had launched their latest submarine from the Mediterranean port of Toulon.

At the start of the 1890s Gustave Zédé, who had been responsible for the success of *Gymnote* after de Lome's death, had persuaded the French Government to construct a new submarine he had designed. At 160 ft (49 m) long and displacing more than 270 tons when completed, she was then the largest submarine boat in the world. Her hull was constructed entirely of bronze and she was more than 12 ft (3.5 m) in the beam, driven by twin electric motors each of which could deliver 360 horsepower and fitted with a 14 inch (35.5 cm) torpedo tube. Zédé died before the workmen at the Mourillon shipyard at Toulon could finish the craft and she was named after her designer.

The *Gustave Zédé* was not a total success. A French naval journal reported: 'The submarine disappointed her supporters. It recently made trials at 8 knots, with several plunges, and the discovery was made that modifications were necessary.

In a general manner of speaking, it is established that the boat is too large to serve any real purpose in war. Moreover, at the moment of descent, its inclination is sometimes so great that her screw emerges, and, meeting with no water resistance, revolves very rapidly. Under such circumstances, in the narrow confines allotted, it becomes very difficult for the men to remain upright, and the value of the boat as a fighting craft becomes very questionable.'

John Holland, however, knew only that the French were forging ahead with *Gustave Zédé* while he was having to negotiate through official bureaucracy, rampant amateurism and Washington politics, all of which were about to create even further delays. In the spring of 1894 he had become manager of the Holland Torpedo Boat Company, at a salary of $50 per month, to oversee the building of his new boat but at the end of the year it was no further forward. It was the following March before the contract was finally signed. Holland had been waiting seven years. By August 1895, his submarine, which the US Government was calling the *Plunger*, was taking shape at the Columbian Iron Works and Dry Dock Company on Locust Point in Baltimore harbour.

At 85 ft (26 m) long the *Plunger* was to be a huge craft. Although she would be only half the overall size of the *Gustave Zédé* she was almost as broad in the beam. She was made of ½ inch (1.25 cm) oil-tempered steel and, to keep to the navy's specifications, was powered by two engines capable of generating more than 1600 horsepower to drive her at 15 knots on the surface. Holland's major innovation was to provide the craft with two power sources because the *Plunger* also had an electric motor capable of generating 70 horsepower to drive the craft at 8 knots when submerged. It was a breakthrough that would change the submarine from a limited experiment into a free and independent ship of war.

Yet the *Plunger* was to frustrate Holland for a further three years. As the boat took shape he began to realize that the specifications imposed by the navy could not be met. Nor did he agree with their performance requirements. Moreover, he was irritated by his lack of control over even minor changes of design. By 1896 he had had enough of the *Plunger* and US navy rules and regulations. He had been working on another design while completing the *Plunger* and felt he could not afford to delay its construction. The company which bore his name also believed that its future should not rest on the *Plunger* alone and agreed to fund and build his new design.

Throughout all the submarine building activity over the last twenty years of the century the British had been watching carefully and their agents in France and America had been sending reports back to London. The British Government had adopted a pragmatic policy of 'wait and see' so far as submarine development was concerned. They felt that as long as they knew the strength of these other nations

His own dreams frustrated by bureaucracy and politics, Holland watched impotent as the French claimed *Gustave Zédé*, their new creation, as the largest submarine boat in the world.

it would not be too difficult to design and launch their own craft when circumstances required. The Foreign Office knew about the American competition for a submarine torpedo-boat and regular reports from France kept them up to date on developments in submarine design.

While the French submarine threat was limited the British were content to sit back and wait. It was only the success of the revolutionary *Narval* which eventually forced them to change their policy. In 1898 M. C. Lockroy, the French minister for the navy, announced a competition to design a submarine of 200 tons which could travel 100 miles (160 km) on the surface and 10 miles (16 km) submerged. Of the twenty-nine entries from all over the world the winner was *Narval*, conceived by Maxime Laubeuf, an inventor once employed by the French navy as its engineer-in-chief.

The shape of the hull that Laubeuf designed indicated that he intended the craft should spend most of her time on the surface and run under water only when absolutely necessary. But that same hull incorporated a feature that was to dominate submarine design up to modern times.

In every previous submarine the hull had been made up of several compartments. Some could be flooded to allow the craft to sink under the surface. Others held only air which provided the safety element of positive buoyancy. Yet another was the control room inside which the crew worked and equipment could be stored. The outer hull of Laubeuf's *Narval,* however, enclosed an inner, pressurized, hull inside which the men controlled the machinery. Between the two hulls there were chambers, some of which could be flooded to dive and then be evacuated by compressed air when the craft needed to surface. Fuel could be stored safely inside others. It was a revolutionary development and gave the French an edge in submarine technology.

By 1898 *Narval* was in the water and successfully completed her trials the following year. French strategists saw the craft with her double-hull construction as the prototype for a range of boats that would not only strengthen their coastal operations but would also operate in the open sea. One of her drawbacks was the time needed to shut down the steam unit before diving, but this was not considered too much of a problem for a coastal defence role. On the other hand, the craft had four torpedo-launchers. Overall, the *Narval* convinced the French nation that it once again led the world in submarine design. In America, John Holland's new boat, the *Holland VI* was still waiting to be given a seal of approval by the US Navy Department.

More than a year before, in the early months of 1897, the craft had been nearing completion at Nixon's Crescent Shipyard at Elizabethport, New Jersey. But there were setbacks and it wasn't until 24 February 1989 that Holland finally took the controls of the boat off Staten Island, New York City, and glided across the surface of Princess Bay. It was the first time he had taken control of one of his own designs since the *Fenian Ram* had been hijacked by his financial backers nearly twenty years earlier.

His masterpiece was 54 ft (16.5 m) long and just over 10 ft (3 m) wide amidships, a hull configuration designed to give a very good hydrodynamic shape for travelling under water. Her Otto gasoline engine could drive her along the surface at 8 knots and a 50 horsepower electric engine gave her a submerged speed of perhaps half that. The engine could be switched from driving the propeller to charging the electric battery – a device all future designers of

OVERLEAF Holland's greatest triumph, *Holland VI*, dwarfed by a Russian warship at the New York navy yard in 1901.

submarines were to copy. She was fitted with a single torpedo tube and would carry two spares. Finally, on 17 March 1898 – St Patrick's Day – Holland took the submarine through her paces and shortly afterwards completed further tests in front of an official observer from the US Navy Department. He had not needed the luck of the Irish. The observer reported to Washington that the submarine had fully proved her ability to propel herself, to dive, to come up, admit water to her ballast tanks and to eject it again without difficulty. He continued: 'I report my belief that the *Holland* is a successful and veritable submarine torpedo-boat, capable of making an attack on an enemy unseen and undetectable, and that, therefore, she is an Engine of Warfare of terrible potency which the Government must necessarily adopt in its Service.'

Despite this report it was another eighteen months before the navy's Board of Construction was satisfied that *Holland VI* had successfully met all the requirements stipulated. Throughout this long period of waiting, John Holland could hardly have been unaware of the immense success of the *Narval* in France.

By Christmas 1899, *Holland VI* was tied up at the Washington navy yard waiting for official acceptance, but its designer was to endure a further four months of tests before the United States' Government finally agreed to purchase the craft on 11 April 1900. Three months later, the Naval Appropriation Act provided for the construction of five boats of the improved Holland type – a figure that was increased to six a few months later. These seven boats were to become the Adder class of submarine: boats *A1* to *A7*. The submarine fleet of the US navy had been born.

Across the Atlantic, however, Maxime Laubeuf was about to play a master-stroke. In 1892 the German inventor Rudolf Diesel had perfected the compression-ignition engine which could run on oil fuel rather than petrol. The safer flashpoint of oil made it more acceptable for use in enclosed spaces and the Diesel engine was also more efficient. Laubeuf recognized early on that *Narval*'s steam-driven engine limited the craft's sphere of operation and within two years of its launch he had designed a modified power unit to drive *Narval*'s successor by a diesel engine on the surface and an electric motor under water. Holland had considered this in 1899 but had not followed it up. Laubeuf called his new boat *Aigrette* and patented his design in 1900.

This innovation would not have escaped the eye of an envious John Holland who had designed what was arguably the best submarine boat in the world, but who had lost control of his firm to one of his partners at the very moment when he could have expected to reap the rewards of his years of pioneering work.

Nor would the French breakthrough have escaped the attention of the greatest naval power in the world, just across the English Channel. The strategists in the

Royal Navy were taking a great interest in events in both France and the United States. In January 1898 the *Gustave Zédé*, on naval manoeuvres, had torpedoed the French battleship *Magenta* while it was at anchor. The attack shocked the majority of French naval officers. It also shocked the Foreign Office and, when they heard of it, the Admiralty. The British Ambassador in Paris warned London: '... belief in the success of the invention is very likely to encourage Frenchmen to regard their naval inferiority to England as by no means so great as it is considered to be in the latter country'. A report from the naval attaché Captain Henry Jackson, added: 'These submersible vessels have now reached a practical stage in modern warfare and will have to be reckoned with, and met, in future European war. One of the most important results of the trials had been to demonstrate that a vessel of this type ... is capable of crossing and recrossing the English Channel from Cherbourg to Portland unaided ... This fact is carefully hid from the public by the authorities, though considered the greatest triumph of this new vessel.'

The Lords of the Admiralty suddenly saw the underwater craft as a very real threat to the British surface fleet. In May 1900 Admiral George Goschen, First Lord of the Admiralty, recorded: 'I have read the whole of the papers most carefully, they are not pleasant reading for clearly great strides are being made in the submarine boat.' The British navy could not allow any foreign rival – particularly the French – the undisputed lead in any type of warship construction. It was time to act and in December that year, Admiral John Fisher argued that, '... we cannot afford any foreign power to possess any type of war vessel superior to our own'. Within months the Admiralty had moved to redress the balance.

The British decided to purchase a handful of submarines so that the Royal Navy could familiarize itself with this new weapon of war. Admiral Kerr, First Sea Lord, argued that '... in doing this I think that we have not only adopted the best course that was open to us, but also done all that we can prudently do. While we are bound to follow up the development of the submarine boats, and thus have at our disposal whatever advantages they may possess, it is not desirable to plunge too heavily as it must first be in the dark, not until experience points us in the direction in which we should work.'

One thing was certain: the French would not sell submarines to the British. The Admiralty therefore looked to the United States.

In December 1900 they ordered five boats of Holland's latest design from the firm of Vickers who were to build them on licence from Isaac Rice. The new head of the Holland Torpedo Boat Company, Rice had effectively squeezed Holland out of the firm and out of all claims for rights to patent royalties on *Holland VI*. HM Submarine No. 1 was launched at Barrow-in-Furness on 2 October

HMS *Aboukir* accompanies the British submarine *D-1* on manoeuvres shortly before the First World War. A few months later she was at the bottom of the North Sea, sunk by a torpedo.

1901 and began sea trials six months later. After a century of watching other nations design and develop, fail and succeed, the British Government, which had at one stage seemed likely to be left behind, suddenly found itself ahead. It had an embryo fleet of the latest design with which it could experiment; and any changes and improvements it wished to make would be at the very forefront of submarine technology.

In France, ironically, the years of submarine innovation were over as an absence of political will coincided with industrial inefficiency to contribute to a general malaise. By 1905 the British had incorporated Maxime Laubeuf's ideas into a diesel-electric power unit to drive their latest model, the *A-13*, for which Vickers had designed a fuel-injection system. Three years later they produced the new D class submarine with twin diesel engines and by 1911 were ready to launch the first of the improved E Class. But a frantic submarine construction programme was also being carried out in Germany.

The French malaise had forced some of the newer designers to look for other markets. One of these was a young Spaniard, Robert d'Equevilley, who had been trying to interest the French navy in a new design. Turned down in France, he immediately looked towards Germany. It was 1902 and he was about to walk, unsuspecting, into a highly charged climate of naval politics and strategy.

Admiral Alfred von Tirpitz, the grand strategist of the German high seas fleet between 1897 and 1915, believed that he needed battleships and other capital ships on a grand scale in order to challenge the British grand fleet. Although the German navy knew about developments in France and America during the last fifteen years of the nineteenth century, he was not interested in submarines. Challenged on this policy in the face of advances in France, von Tirpitz replied: 'The submarine is, at present, of no great value in war at sea.' In the Reichstag, in 1901, he argued that the configuration of Germany's coasts and the geographical situation of her ports meant that she had absolutely no need of submarines, which he considered to be purely defensive weapons. He constantly spelt out every imperfection in French submarine design to reinforce his message that the German navy should not become involved with technology still under development. 'We have no money to waste on experimental vessels,' he said during another speech. 'We must leave such luxuries to wealthier states like France and England.'

His thinking dominated German naval development until after the battle of Jutland in 1916: only after great resistance did he agree to allow limited expenditure on submarine development in the years leading up to the First World War.

It was into this environment that d'Equevilley took his plans for a new submarine in 1902. Turned down by the German navy the young Spaniard eventually took his design to the Friedrich Krupp engineering firm in Essen.

In February 1902 Krupp had bought the Germania yard at Kiel and plans for the first experimental submarine, *Forel*, were drawn up. She was laid down in July that year but represented no great advance on the French *Gymnote* which had been completed sixteen years before.

Forel's one great defect was that she had to be electrically charged from the shore. Krupp's engineers were determined that their next design would not be so limited and experimented with a combination of internal combustion engines, dynamos and electric motors. They also decided that fuel with a high flashpoint should not be carried inside the hull. D'Equevilley's plans bore an amazing resemblance to those of Maxime Laubeuf. This was unsurprising: d'Equevilley had worked as Laubeuf's assistant before heading for Germany. In 1904 d'Equevilley took out a patent on a double-hulled craft, which resembled Laubeuf's design for *Narval*. Three boats that Russia ordered from Krupps for her 1904–5 war with Japan were built on this principle, powered by twin heavy-oil engines made by the firm of Korting Brothers. These Karp class boats represented the real beginnings of German submarine design; the first craft built for the German navy was an only slightly improved version. When it was launched on 16 April 1906 it displaced 238 tons and was 139 ft (42.5 m) in length. Powered by Korting engines that burnt crude paraffin, she could travel at nearly 9 knots when submerged, only 2 knots faster on the surface, and had a range of about 2000 miles (3220 km). She had a single torpedo tube in her bows and carried three 18 inch (45.5 cm) torpedoes. Although the craft was not a great success, she bore the code number U1. One report on her trials recorded: 'The U1, commanded by Lieutenant Bartenbach, has undergone trials in the North Sea with a complement of nineteen men. Her small displacement renders this boat unfit for operations at any distance from the coast, as observations taken showed that her employment in the high seas is attended with danger.'

In 1908 Germany also took the U2, which was little more than a replication of the U1, into her fleet. The French at that time had sixty submarines, the British only slightly fewer and the Americans about twelve. However, many of the boats commissioned by all three countries were small and feeble, fit only for coastal defence. Within two years Krupp had completed their eighteenth U-boat. The nineteenth, U19, built in 1910, was fitted with twin six-cylinder diesel-electric motors which provided 1700 horsepower to drive her at 15 knots on the surface.

Eight years earlier Germany had not owned a single submarine. Now she was preparing a fleet of the fastest and most powerful submarines in the world. The long-held beliefs about naval strategy held by John Fisher, First Sea Lord of the Admiralty and disciple of the *Dreadnought*, and his opponent Admiral Alfred von Tirpitz, chief architect of the German navy's reliance on huge capital ships, were about to be shattered.

When the *U19* was launched the inevitable conflict between Great Britain, at the height of her imperial and naval greatness, and Germany, nursing her own dreams of empire, was no more than four years away.

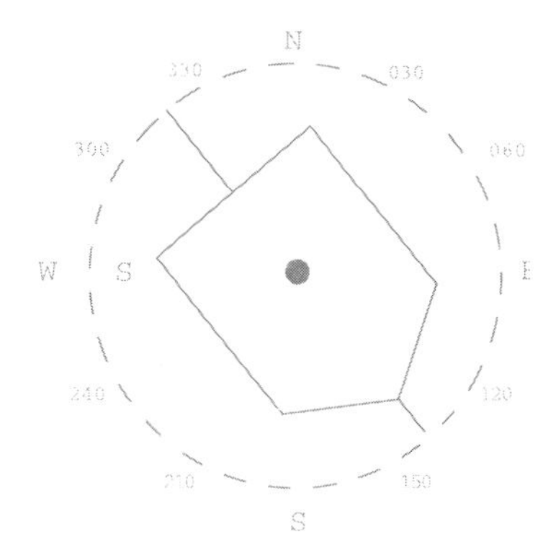

CHAPTER 4

THE ULTIMATE TEST

In late October 1916, Sir John Jellicoe, Commander of the Grand Fleet, was a worried man. He had just despatched a letter to the Admiralty in which he had stressed the need for Great Britain to employ new methods in the submarine war being waged against Germany. By the early summer of 1917, he had written, supplies of food and other necessities might be reduced to such an extent that Britain would have to seek peace without victory. The German submarine presented, he argued, 'the most serious menace with which the Empire has ever been faced'. The Admiralty, in turn, indicated Jellicoe's concerns to the Government. 'No conclusive answer has as yet been found to this form of warfare,' they warned, 'and perhaps no conclusive answer will be found.'

From the first days of the First World War the U-boats had dominated the seas around Great Britain. While the main British strategy was to blockade the English Channel and the approaches to Germany's ports in an attempt to cut off food supplies to the German heartland, the U-boats were seemingly free to roam at will, sinking shipping at all points of the compass around the British Isles.

When Jellicoe wrote his letter to the Admiralty the Germans had temporarily suspended their policy of unrestricted warfare on the high seas which had led to one of their submarines sinking the *Lusitania:* more than 1000 people on the British liner, 128 of them US citizens, had died. But the British Government knew that it was only a matter of time before the U-boats once again began to target the grain ships and merchantmen that delivered food and supplies to Britain from the United States and from her colonies around the world. It also knew that the impact of the British blockade, by then almost two years old, meant that the German people were finding it increasingly hard to find food and that young and old, healthy and unhealthy, were suffering – and that the only way for the enemy to break the blockade and get sufficient food into Germany again might be to force the British to sue for peace by inflicting the same hunger and despair on the people of Great Britain.

The winter of the First World War at sea was the harsh, long, unremitting, cruel and desperate eighteen months of U-boat attacks which threatened to bring Britain to the point of starvation by the middle of 1918. In its spring, four years before, the war had begun with a mixture of hope and high spirits; the pride and national respect of two mighty nations were invested in the huge fleets of high-sided battleships that rode at anchor in naval bases from Kiel to Plymouth. Few people had taken much more than a passing interest in the flotillas of tiny submarine craft that had been slowly built up since the turn of the century.

Spring: AUGUST 1914 – APRIL 1915

From the bridge of the cruiser HMS *Birmingham* the shape was unmistakable. A U-boat, hove to on the surface some distance away in the thin sea mists, was obviously in some difficulty. The British sailors could clearly hear the sounds of hammering across the stretch of water which separated them, even above the noise of the cruiser's engines.

Aboard the *U15* one of the look-outs must have seen the British warship and triggered the order to dive, for, even before the cruiser turned sharply towards the disabled submarine, the U-boat slowly began to move ahead. The order, shouted by word of mouth – electric diving alarms had been disapproved on several occasions on grounds of their high cost – was futile. The craft, one of the first built for service in the Kaiser's navy, could not submerge in less than five minutes and long before then the cruiser was on her. At speed the *Birmingham* rammed the submarine and sliced her in two.

The *U15* was the first U-boat to be sunk in the war and her vulnerability that morning led some British politicians to believe that U-boats would not be the threat many had suggested. Indeed, two of the twelve U-boats sent to patrol the North Sea the day after war was declared failed to return. Apart from the *U15*, the *U13* was never heard of again.

John Holland died at his home in New Jersey five days after the sinking of the *U15*, convinced that the submarine was about to change the face of naval warfare. Within a month his prophecy began to come true, with such force that submarines dominated naval action for the rest of the war. But it was the skill and ability of individual U-boat captains and some British commanders, and not the demands of politicians, that convinced governments that submarines had a much more important role to play than even their most faithful supporters had believed.

Reports of the sinking of the *U15* got back to other U-boat captains. One of them was Otto Weddigen, who had brought his boat limping back to port from an early patrol after a technical breakdown. 'Our first submarine advance resulted in no

Kapitänleutnant Otto Weddigen. In less than two hours, a well-trained crew, his sheer determination and three unsuspecting British cruisers combined to change naval strategy for the rest of the war.

damage to the enemy, and we had lost two boats out of twelve,' Weddigen's First Officer, Johannes Spiess, recalled later. He added that all they could do at the time was grit their teeth and await a better chance while vowing to remember the *Birmingham* and exact vengeance for the *U15*.

That honour fell to Kapitänleutnant Otto Hersing. He was the commander of the *U21*, only the third diesel-electric boat to be launched, and was to become one of the most famous U-boat commanders of the war. On 5 September 1914, in mountainous seas which helped hide his periscope from his target, Hersing saw a British cruiser heading towards him. The warship was making no more than 6 knots. He stole to a point close to the path of the oncoming warship and lay in ambush there before loosing a single torpedo which struck HMS *Pathfinder* at the waterline just under the forward funnel. The entire forepart of the vessel was destroyed and the stern rose to stand vertical in the air before the vessel disappeared under the waves. It struck Hersing that the whole affair, from the moment of releasing the torpedo to the disappearance of the *Pathfinder*, was only three minutes. Fewer than half the cruiser's crew of 360 men were saved and only one lifeboat managed to get away before the *Pathfinder* went down. All the other survivors were found clinging to wreckage. Without direct radio communication with its base, the *U21* had no means of getting the news back to Germany. When word did trickle in days later, Hersing became the first submarine ace of the war.

A little more than two weeks later his friend, Otto Weddigen, reinforced the message that John Holland was right and that slow-moving capital ships could no longer steam imperiously around the North Sea without risking complete destruction.

Kapitänleutnant Weddigen had under his command a far older boat than Hersing's *U21* when he was ordered into a position off the island of Svenigen on 20 September to attack British war vessels. The Germans believed that ships might try to land British troops on the Belgian coast to intercept German regiments marching through Belgium to the Marne where fierce fighting was in progress.

Johannes Spiess had been with Weddigen on the *U9* since 1912. Nothing much was expected of U-boats at that time and very little had changed in the way of expectation by the morning of 22 September 1914.

Before taking the *U9* to war Weddigen had practised the skills which he believed would keep himself and his crew alive in battle. Spiess recorded that they exercised diligently: 'Our latest exercise was the reloading of torpedoes at sea both on the surface and submerged ... it is a remarkable coincidence that *U9* was the only German boat that had practised this manoeuvre in time of peace and was two months later to make use of this experience in real earnest against three English armed cruisers. No one thought at the time of such a distant possibility....'

Weddigen demanded not only that loading routines become second nature to his crew, but also that his men become totally versed in salvo-firing so that he could release four torpedoes one after the other from both bow and stern torpedo tubes. It was the first time such synchronization had been attempted in training and it attracted considerable attention at the time. However, the crew were about to suffer because, although the U-boat looked streamlined in the water, Spiess described an entirely different story below decks.

'Inside the pressure hull, which was cylindrical, was the forward torpedo room containing two torpedo tubes and two reserve torpedoes. Further astern was the warrant officers' mess which contained only small bunks for the quartermaster and machinist and was particularly cold.

'Then came the commanding officer's cabin, fitted with only a small bunk and clothes closet, no desk being furnished. Whenever a torpedo had to be loaded forward or the tube prepared for a shot, both the warrant officers' and commanding officer's cabins had to be completely cleared out. Bunks and clothes cupboards then had to be moved into the adjacent wardroom – which was no light task owing to the lack of space in the latter compartment.'

Spiess also described the submarine's tiny bunks, which allowed off-duty watchkeepers to sleep only on their sides – and only in one direction because the cover of an electric fuse box on the aft wall sometimes flew open and a stretched-out foot could cause a short circuit. The stove also short-circuited sometimes and the crew only had hot food if they could cook it on a gasoline stove on deck.' His description of the inside of an early U-boat is an extended picture of extreme discomfort from the start of a trip to its end. But this was the machine which was about to change the course of a war.

Some time before seven o'clock on the morning of 22 September, Weddigen identified three British cruisers on the horizon. He decided to attack. Spiess went below: 'I ordered reserve torpedoes made ready for the manoeuvre of submerged firing, reloading and firing again which we had only a few weeks before accomplished successfully the first time in practise.' Weddigen, who had closed to 550 yds (500 m), gave the first command to fire at 07.20 a.m. Almost exactly half a minute later HMS *Aboukir* turned slowly over to one side and disappeared under the waves. Spiess could not believe what he then saw through the periscope. 'The other two cruisers, companions of the sinking ship, were standing by to take survivors aboard. What a fatal mistake! British warships never did anything like that again during the length of the war. Weddigen made ready for another attack. I hurried to the forward torpedo room. I imagined I was passing through a madhouse. Men were running furiously back and forth, a big group of them. First they rushed forward and then astern. The chief engineer at the depth rudder was helping to keep the boat on an even keel by a process of ballast shifting.' In the very first submarines the discharge of every torpedo temporarily unbalanced the boat's trim and there was no possibility of lining up for another shot until it was stable again.

At 07.55 a.m. the order was given to fire both bow tubes at HMS *Hogue* from a range of only 330 yds (300 m). Weddigen's torpedo hit home again but the U-boat's forward momentum was such that as the *U9* turned away in a circle she almost touched the stricken *Hogue* with her periscope before she was once again stabilized and Weddigen prepared to attack HMS *Cressy*.

This time he decided to fire both stern torpedoes from a range of 1100 yds (1000 m). Almost exactly one hour after the first cruiser had gone down, the third crippled British warship stopped, alone, on the surface of a smooth sea surrounded by hundreds of drowning men.

'It was a long shot at 1000 metres and the victim did not sink,' Spiess said later. 'Weddigen then decided to fire our last torpedo at the damaged ship. At 08.55 a.m. it left the tube and struck the *Cressy* abeam . . . The giant with four funnels fell slowly but surely over to port and, like ants, the crew crawled first over the side and then on to the broad flat keel until they disappeared under the water.'

When news of the sinking of the three British cruisers got back to Germany the Kaiser was reported to be 'in seventh heaven'. The country made ready to fête Weddigen and his crew when they returned and the national newspapers told and retold the latest U-boat success so close on the heels of Hersing's achievement in sinking the *Pathfinder*.

In England the mood was sombre. One week after the triple disaster in the North Sea, Sir John Jellicoe wrote to Winston Churchill, then First Lord of the Admiralty: 'It is suicidal to forego our advantageous position in the big ships by risking them in waters infested with submarines. The result might quite easily be such a weakening of our battle fleet and battle cruiser strength as seriously to jeopardize the future of the country by giving over to the Germans the command of the open seas.'

For the first three months of the war, arguments continued at the highest levels of the Admiralty about the best and safest way to use the battle fleet. Submarine alarms caused it to be moved twice to safer anchorages but vital anti-submarine defences at Scapa Flow in the Orkneys only really began taking shape two months after Weddigen's attack and were not completed until the following summer.

Old merchant ships were sunk to block some channels and booms were laid to stop U-boats entering others. Defensive minefields were laid in the principal entrances to the anchorage and nets, seaplanes, shore batteries and searchlights were installed. Behind this array of defences the British Grand Fleet finally felt secure from the tiny U-boats that had suddenly proved them all so vulnerable.

It had not, however, been all one-way traffic. The first patrol of twelve U-boats had been ordered because the German high seas fleet commanders had no strategic plan of campaign. The British, on the other hand, had a very clear idea of how they wanted their submarines to work and before dawn on the day war was declared two E class submarines were sent to patrol the Heligoland Bight and report back. They were the advance unit of a fleet of submarines which was soon to blockade the North Sea approaches to the Baltic and the ports of north Germany.

The captain of the E-9 was Lieutenant-Commander Max Horton. He was a naval officer with a reputation for gambling and irreverence, but an excellent submarine commander. On the night of 12 September he had taken his boat down to rest on the bottom of the sea for the night. It was a technique that both British and German commanders had perfected. When Horton brought the E-9 up to periscope depth the following morning, 6 miles (10 km) south of Heligoland harbour, he saw a German light cruiser, the *Hela*, breaking through a patch of mist. From a distance of 650 yds (600 m) he fired two torpedoes and sank the German warship with hits amidships. Three weeks later, back on patrol in the same waters, he hunted and sank a German destroyer which,

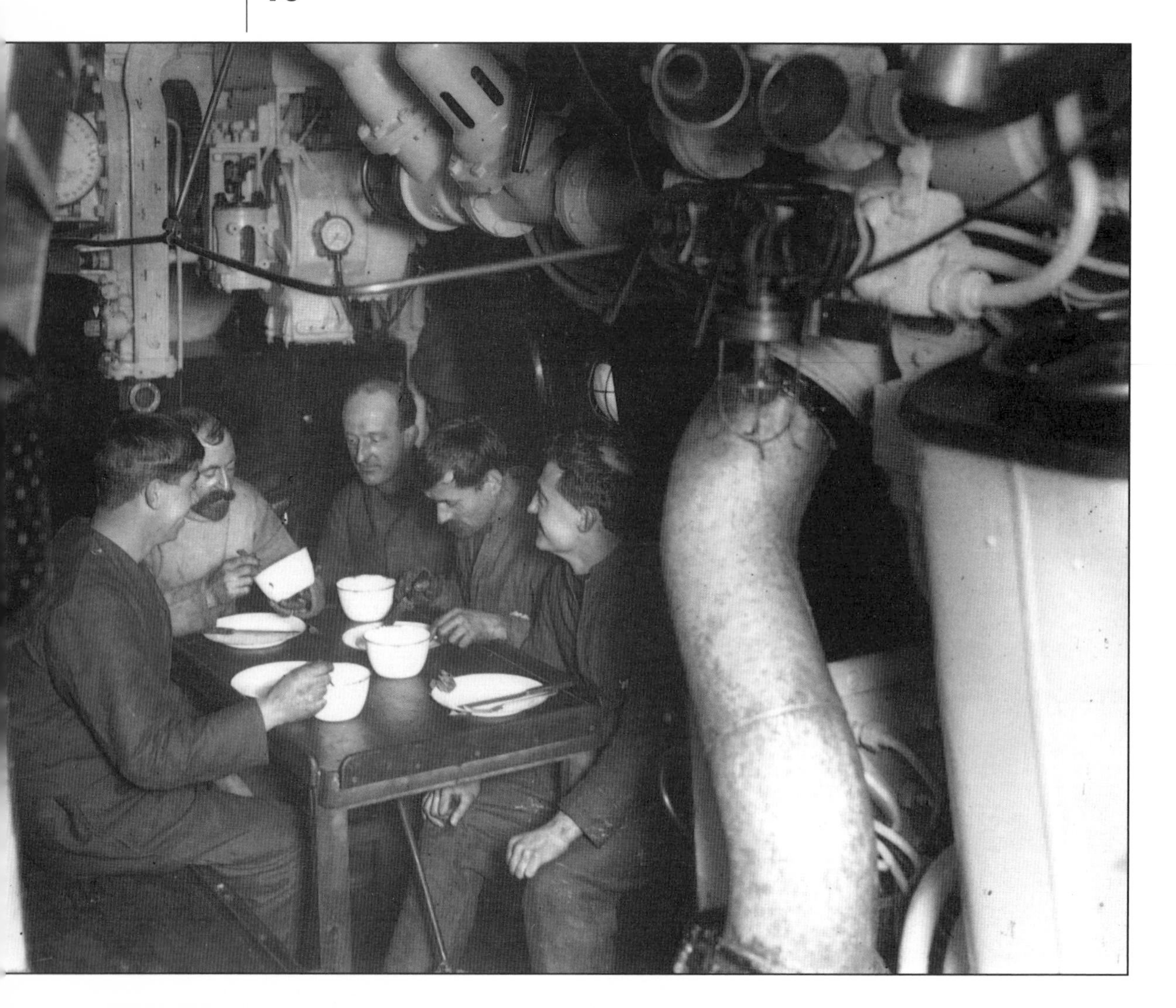

ABOVE The petty officers' mess on a British submarine in the First World War. Bread kept about three days before going mouldy. There was no bath and one lavatory for the three officers and twenty-nine men.

LEFT A First World War British submarine captain taking periscope sightings. Conditions were sometimes so cold in the Baltic that frozen periscope mechanisms rendered submarines blind.

An artist's depiction of activity at the nerve centre of a German submarine during the First World War. The traditional ship's wheels are an interesting hangover; the cleanliness of the uniforms extremely unlikely. Painting by Willy Stöwer.

because of its speed and manoeuvrability, was among the most difficult targets for a submarine. Horton wrote to a friend: 'To hit a destroyer always requires maximum luck. She went up beautifully, and when I had a chance of a good look round about five minutes afterwards, all that was to be seen was about fifteen feet of bow sticking up vertically out of the water.'

Less than a month later Horton, accompanied by Lieutenant-Commander Noel Laurence in the E-1, was on patrol in the Baltic where German battleships and battle cruisers were considered safe from British submarines and from any elements of the British grand fleet they might have met in the North Sea. The submarine commanders were interested in disrupting the passage of merchantmen supplying Germany with goods from Sweden while the Baltic ports were free from ice. Horton and Laurence proved themselves successful on their roving commission in the Baltic. They sank several merchantmen between them and Horton sank his second destroyer.

Horton used the winter of 1914 to experiment with his boat in conditions well below zero. When the ice formed on the sea he took the E-9 on a test trip to see how far it could go. The submarine was soon encased in ice but Horton was anxious to discover whether diving was still possible with slush ice in the craft's vents and valves. He discovered that once the submarine was under the surface of the sea the salt water thawed the slush and left the diving efficiency of the boat unaffected. However, on the surface, the periscope froze in its tube and could not be moved and the ice jammed the caps on the torpedo tubes making them useless.

The British Admiralty was soon informed that German ships could no longer freely move around the Baltic. It had become known as Horton's sea and, to confirm his reputation, the Englishman sank one more German destroyer there before he and Laurence pulled out in the summer of 1915.

Horton and Laurence for the British and Hersing and Weddigen for the Germans proved that it was the skill of the captains and the efficiency of the crew that determined the success of any patrol. These brave men experimented with their submarines and made their craft and crews perform to standards of efficiency, concentration and stamina that only conditions of war can produce.

Levels of skill in helmsmanship, trimming, vane operating, engine maintenance and watchkeeping increased rapidly in conditions where lack of concentration by any member of the crew might result in the sudden death of everyone on board. It was these general skills and the sheer audacity and determination of the most gifted commanders, allied to an element of luck, that made some men famous in their own countries and feared as cold-hearted killers by their enemies. Yet they were successful mainly because they were practical enough to understand and make use of the technical capacities of their vessels; because they were skilful enough to make unexpected situations work for their own benefit; and because they were able to command the respect and absolute trust of their men – which gave them the freedom to take their submarines into the most dangerous of waters to search out and destroy the enemy.

It was partly pure luck that neither side radically transformed the basic design of their submarines throughout all the years of conflict. Since the Germans had come to submarines so reluctantly and the British with blatant opportunism, it was sheer coincidence that the basic design of their boats conformed exactly to the roles they wanted them to play. The different origins of the craft meant that the Holland-inspired British submarines performed more efficiently when submerged while the French-influenced U-boats performed most efficiently on the surface.

However, the stories of life below decks are remarkably similar. The amount of time the submarines could spend submerged was limited. For the British there was a constant anxiety: would the electric batteries run out and force them to surface or would they use up all the oxygen first? When running submerged, men had to tolerate the lack of oxygen caused by rebreathing the limited supply of air. Mild carbon-dioxide poisoning caused them to breathe heavily, slowed their reactions and brought on varying states of depression.

Geoffrey Clough, a radio telegraphist who patrolled the North Sea for the last two years of the war, remembered it vividly: 'After about ten hours submerged, a lighted match would soon fizzle out and refuse to burn, there being so little oxygen in the atmosphere. Had it not been proven under wartime conditions, twenty hours would have been considered beyond the capability of a World War I submarine.'

Desperate for air, but unable to surface for fear of being spotted and attacked, crews in such circumstances were ordered to lie down and breathe as shallowly as possible in order to preserve the remaining oxygen. Conditions were made worse by the steady build-up of air pressure as a result of leaks from the tanks of compressed air. The pressure could become so high that when the boat did finally surface and the hatch was opened any man standing immediately under the hatch was in danger of being shot out of the conning tower like a human cannon-ball. But such dangers were exotic possibilities on most submarines. Geoffrey Clough served only on routine patrols, where tales of bravado were always about other submarines, and recalled details of domestic routine: 'There was little privacy and little comfort in a submarine. There was no bath and only one lavatory for the use of all three officers and twenty-nine men aboard. Few shaved and no one changed their clothes from the beginning to the end of the voyage. The officers used eau-de-cologne to mask their body odour and the

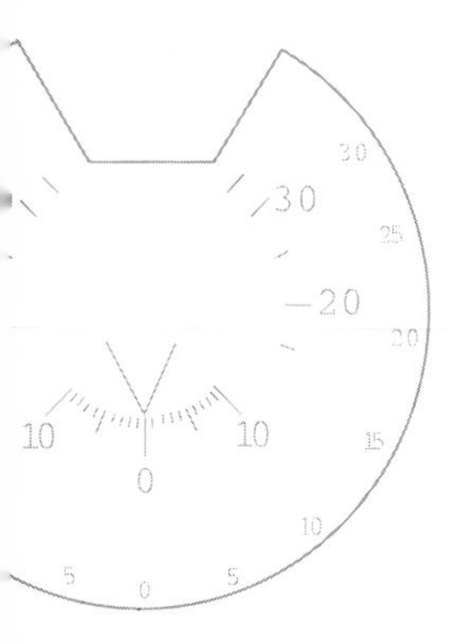

indescribable damp, oil-laden, stale smells of the sweating interior of the boat. But because of this closeness and the shared hardship and danger and because there was no room for men who could not be relied on, a submarine's complement was a uniquely tight brotherhood "One for all and all for one" as it was expressed, we were like a great family isolated on the wastes of the oceans.'

There were often long stretches of patrol when the men had little to do. Smoking was forbidden when the boat was submerged and there was little room in which to move around. The quality of the food was often poor; the crew lived mostly off tinned and dry food. Bread would keep for about three days before turning green on the outside and potatoes for just a little longer. Some men remembered that in the increasingly stale, fetid atmosphere such deprivations did not seem a particular hardship.

Geoffrey Clough remembered many of these difficulties being compounded by the terrible cold experienced by British submariners who patrolled the North Sea. 'In winter off the Danish or Norwegian coasts, spray would turn to ice immediately on the aerials, conning tower and watchkeepers alike. Ice would often have to be kicked off the canvas bridge-screen before it could be folded and stowed before diving. Look-outs were secured to the periscope standards by lifelines to prevent them being lost overboard. Even when you went below soaked to the skin and miserably cold, there was not much comfort to be obtained since your clothing was permanently damp from condensation. The inside of a steel cylinder gets pretty cold in winter. Ice on coats might not melt for some hours after diving, when hung up below.'

But above all he remembered the deadly monotony of his wartime patrols: 'There was so little opportunity for action with the enemy. We were the eyes of the fleet, watching and waiting for the German capital ships to put to sea or to catch a U-boat transitting our area. There was very little merchant shipping getting through our blockade to German ports. Occasionally naval intelligence was able to alert us to a U-boat known to be passing within range of our patrol billet and if we were lucky we might catch sight of him, but opportunities for putting in a successful attack were few. It was therefore even more vital that a crew's operational efficiency was maintained at a high peak in order not to miss the few chances that came our way.'

But for Geoffrey Clough, and so many others, opportunities to engage in action were rare. For every crewman who adventured with Weddigen and Horton, Hersing and Laurence there were thousands who froze and ached and were sick and tired in boats that never fired a shot in anger throughout the entire war. And if they found themselves in a position where they might have attacked, the Admiralty sometimes ordered otherwise.

In the early years of the war, British submarine commanders who sighted an enemy merchant ship during their relentless vigil were not allowed to sink her; they were restricted to reporting these ships back to base. Britain's Liberal Government insisted on applying the rules of international warfare until 1916 and British ships were forbidden to fire at unarmed merchant vessels.

For Horton and Laurence the spring of the war was ending but for Kapitänleutnant Otto Hersing in the *U21* and Lieutenant-Commander Martin Nasmith in the *E-11*, the chance to discover even more about the capabilities of the craft they commanded was about to begin.

Summer: FEBRUARY 1915 – MARCH 1916

In November 1914 the Allied campaign against Turkey had begun. The British Government had decided that rather than simply continuing to slog away at entrenched German lines in northern France, it would take the initiative by moving an army through the Balkans in an attempt to turn the German positions in Belgium and France. By early January 1915 Russian positions in the Caucasus were under threat from Turkish forces whose leader, Enver Pasha, had signed an alliance with Germany. The British promised the Russians a relieving action against the Turks and in February 1915 ships of the Mediterranean fleet were ordered to bombard Turkish positions on the Gallipoli peninsula. The plan was for British forces to capture the western shore of the Dardanelles and take Constantinople.

Towards the end of April, after his achievements in the North Sea in the opening months of the war, Hersing was ordered to undertake a mission which no U-boat had ever attempted before. He was to take the *U21* from Wilhelmshaven naval base to Constantinople to help the Turks repel the British naval attacks on the Dardanelles. But his craft could not carry enough fuel to make the journey without stopping and there was no friendly port between the North Sea and the Turkish capital. Arrangements were made for one of Germany's Atlantic steamers, the *Marzala*, to rendezvous with the *U21* off the coast of Spain. By the time Hersing was ready to leave on 25 April 1915 the English Channel was almost completely blocked with barriers of nets and mines and he therefore headed for the northern tip of the Orkneys before turning south for Cape Finisterre, the north-western point of the Spanish coast, where he planned to rendezvous with the steamer. The journey there took seven days. The *U21* passed Gibraltar on 6 May and less than three weeks later, in the early hours of 24 May, Hersing and his crew were waiting for daybreak and the chance of attacking the British battleships off the Dardanelles. Hersing's simple idea was surprise. The British would never expect a German U-boat to attack them in the Mediterranean.

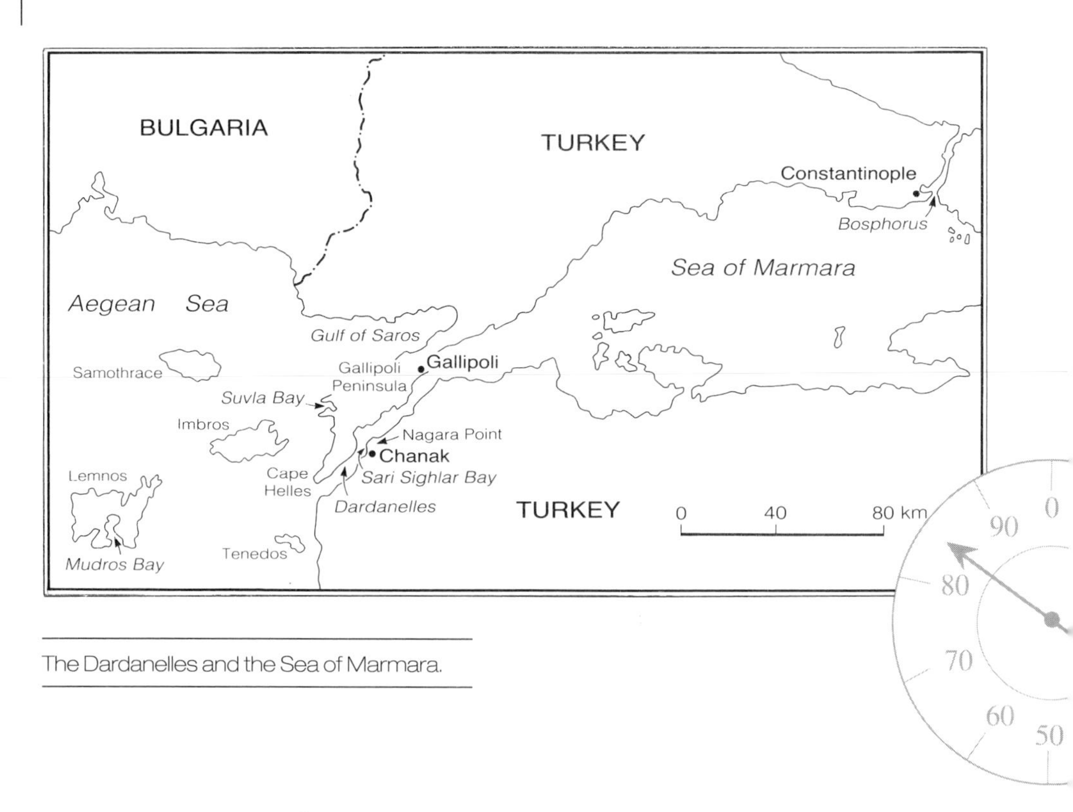

The Dardanelles and the Sea of Marmara.

On that morning of 6 May Hersing dived to 70 ft (21 m) and headed towards a British battleship, passing far below its scurrying defensive shield of destroyers and other patrol craft. He could clearly hear the steady beat of their propellers above him as he began the undersea manoeuvres that would give him the best target for the torpedoes waiting in his forward tubes. He came up to periscope depth, indicating his position for only the briefest of moments to any alert patrol boat, while he took in the scene.

Some 300 yds (275 m) in front of him HMS *Triumph*, one of three giant battleships of the Majestic class, blasted away with salvoes from her heavy guns at the Turkish positions among the hills overlooking the Dardanelles.

Hersing thought that no undersea craft had ever been offered such a target. He gave the order to fire the torpedo. In his excitement, he left the periscope up instead of diving away from the scene and was able to see the torpedo – as a streak of white water – heading towards the bow of the unsuspecting warship. As he was watching the white trail of that first torpedo fired in the Mediterranean, he suddenly realized how vulnerable he was. British destroyers were bearing down on the *U21* at high speed and Hersing knew that unless he did something quickly his boat would be sunk. He ordered full speed ahead and followed the course of his torpedo and dived under the sinking battleship. The destroyers whined overhead as they homed in on the spot where the wake of the torpedo had begun. But Hersing, already beginning to use the submarine in ways never conceived in theory, was by then safely on the other side of the stricken battleship and moving steadily away from the scene.

Hundreds of British craft were searching for him and Hersing had no intention of surfacing that afternoon or evening. He waited until darkness, when the British boats should have been called off. The submarine had been under water since dawn the previous morning. Inside the boat the air had become so foul his crew could hardly breathe. They were drowsy, their limbs felt heavy and moving about inside the cramped quarters took an impossible amount of effort. All this was remedied in only a few brief minutes once the pure, cool air of the Mediterranean night flowed into the craft as they surfaced and began to recharge its batteries. Hersing circled back to the site of the sinking of the *Triumph*, but it was mid-morning the next day, off Cape Helles at the tip of the Gallipoli peninsula, before he picked up the scent again. On the beaches several large transport craft were landing troops but at anchor only 500 yds (460 m) offshore stood a battleship one-third as big again as the *Triumph*.

HMS *Majestic* was covering the landings and was surrounded by a cordon of small boats that acted as a live and active screen to forestall any torpedo attack. The sinking of the *Majestic*'s sister ship had obviously had its effect on those responsible for the rest of the fleet. Hersing was worried that any torpedo strike might be intercepted by one of the small boats moving between him and his target across the choppy sea. Then from 600 yds (550 m) he eventually saw a gap and gave the order to fire. This time he dived at once and moved away from the starting-point of any tell-tale wake in the water. He heard the explosion as the torpedo hit home, looked through the periscope to check that the *Majestic* was fatally damaged and then dived away from the attacking destroyers. Half an hour later he saw through the periscope that the battleship had turned over completely. Only her keel remained above the surface of the sea.

Within a few hours a third battleship, HMS *Queen Elizabeth*, was removed from the war zone and sent home to England for safety. It was no longer possible to deny that the presence of U-boats in the Mediterranean had had an impact on the ability of the British navy to support the land offensive in the Dardanelles. Even before Hersing had sunk the two battleships, the mere presence of his U-boat in the area (he had been sighted shortly after negotiating the straits of Gibraltar) had caused the Admiralty to authorize the withdrawal of half the heavy ships in the Mediterranean to the safety of the island of Imbros. The author Compton Mackenzie, then serving with Naval Intelligence, commented on hearing of the decision: 'It is certain that the Royal Navy has never executed a more demoralizing manoeuvre in the whole of its history.' It is not recorded what he said after the sinking of the two battleships.

For Hersing, it was enough that one small craft had sunk two battleships and driven away England's flagship during a critical battle and saved the Turks from the bombardment they had been enduring from the massive 16 inch (40 cm) batteries of the capital ships positioned just offshore. He patrolled the Dardanelles long enough to ensure that the battleships were not replaced and then, on 5 June, he berthed at Constantinople where he found that the British had put a price of £100 000 on his head. It had taken him only four days to travel from the entrance to the Dardanelles through a pre-determined opening left for him in the dangerous maze of mines and nets with which the Turks guarded the entrance to the Sea of Marmara.

These defences provided the biggest challenge of the war to British submarines. Three British B class craft, later reinforced by more modern French and British squadrons, had been posted off the Turkish coast as part of the Dardanelles naval operations. They were ordered to attack enemy shipping so that the Allied advance to Constantinople would be easier. The task facing them was enormously difficult. The Marmara is a deep inland sea some 100 miles (160 km) long and 50 miles (80 km) wide. It is small enough for a submarine to patrol from one end to the other in 24 hours – and for it to be visible from the shore in clear weather. In those early days, when submarines could not stay submerged indefinitely, the boats had to risk sitting on the surface to recharge their batteries.

The Straits of Chanak, the gateway to the Marmara, are some 35 miles (56 km) long and less than 1 mile (1.6 km) wide at one point. The British had little navigational information, but knew that there was a treacherous current that could swing a submarine off course and even throw it on to sandbanks where it would be stranded and at the mercy of the enemy. British naval charts showed a strong current flowing from the Marmara into the Mediterranean. Any submarine attempting to travel up the straits against the current would therefore risk draining its batteries completely. Yet it was inconceivable that a submarine should surface in the straits. The nearby shore was heavily defended. Rows of powerful searchlights could pick out any target and were backed by shore-based artillery and torpedo tubes, and patrol craft and destroyers in the straits themselves. A submarine risked destruction even by raising its periscope to check position. A craft that surfaced to recharge her batteries would have been destroyed immediately. To add spice, naval intelligence indicated that the Turks had sown the passage with ten successive rows of mines at depths varying between 16 ft (5 m) and 30 ft (9 m). The threat of British submarines moving from the Mediterranean into the Marmara had already been foreseen.

A German submarine depicted leaving Istanbul and the Sea of Marmara for active service in the Mediterranean where Lothar von Arnauld de la Perière, captain of the *U35*, was to become the leading U-boat ace of the war.

The British Admiralty, recognizing the difficulties, offered the challenge to any submarine commander willing to take it. The first to succeed was Lieutenant Norman Holbrook in the B-11, an old petrol-engined boat which had been fitted out with new batteries. On Sunday, 11 December 1914, these gave her the strength to penetrate 12 miles (19 km) up the straits as far as Sari Sighlar Bay where an old Turkish battleship, the *Messoudieh*, was anchored. Holbrook hit her with the first torpedo he had ever fired. Then, as shells from every gun in the bay hailed down on him, he headed back to the Mediterranean. During the next eight, perilous hours he ran aground on several sandbanks before getting his exhausted crew back to safety with the last spark of power from the batteries. For his efforts, Holbrook was awarded the Victoria Cross – the first ever to be won by a British submariner.

The appalling casualties suffered by British forces at Gallipoli forced the Admiralty to try any tactic to ease the situation on land. The Marmara was the

Turks' most important supply route to the Dardanelles. Supplies transported overland had to take a much longer route via incomplete railways on the northern and southern shores of the sea or along slow, rough roads. An allied submarine in the Marmara could attack enemy shipping and create havoc among the ships taking supplies and troop reinforcements to the Turkish front lines.

The French submarines *Saphir* and *Joule* and the E-15 all attempted the passage through the straits but each in turn was lost. However, an Australian submarine, the *AE*-2, managed to get into the sea on the morning of 26 April 1915, the day after the first landings had been made on the beaches of Gallipoli. When the signal that the *AE*-2 was in the Marmara arrived, the British naval command, based at Mudros harbour on the island of Lemnos, was discussing the first reports of the horrific casualties suffered by Australian and New Zealand troops on the peninsula. Admiral Sir Roger Keyes read the signal aloud to the other officers present and added: 'It is an omen – an Australian submarine has done the finest feat in submarine history and is going to torpedo all the ships bringing reinforcements, supplies and ammunition to Gallipoli.'

Keyes' statement was tragically misleading. The *AE*-2 was not enjoying great success in the Marmara. Four days into his patrol Lieutenant-Commander Henry Stoker had attacked several ships, including a battleship, but his torpedoes had run too deep and missed. On the fifth, the *AE*-2 was holed by shells from a Turkish gunboat and Stoker was forced to scuttle his craft. Admiral Keyes' hopes now rested on the E-14 commanded by Lieutenant-Commander Edward Boyle, which had also negotiated its way through the straits. Boyle's one-man blockade of the Marmara was more successful. In one attack he sank a 5000 ton troopship with 6000 troops and an artillery battery aboard. Overall, however, his effectiveness was limited. After twenty days on patrol he was recalled and Admiral Keyes ordered Lieutenant-Commander Martin Nasmith into the Marmara until Boyle and the E-14 returned from a refit in Malta.

Nasmith, an experienced and ambitious officer of whom the Admiralty had high hopes, flew up the straits in a small two-seater plane to survey the route and fix navigational points. He knew that when he travelled the unknown passage submerged he would only have seconds to raise the periscope, look around and work out the E-11's location. He noted the positions of the harbours and lighthouses as well as the contours of the mountains. He re-studied the navigation maps in the hope of understanding the complicated pattern of currents in the straits and trained his crew to meet every possible danger. He practised long dives, keeping the E-11 submerged for twelve hours until her oxygen and batteries were almost exhausted and took the boat far below the official safety limits to show the crew that they could safely dive below the mines. He practised crash dives and manoeuvres in the darkness and loaded the boat with twelve torpedoes – the maximum it could carry.

Nasmith also fitted a jumping wire from the bow to the top of the periscope and then to the stern to deflect the cables from mines, and had specially designed guards fitted over propellers, hydroplanes, bow caps and all other projections that might foul an obstruction. In addition, the E-11 was fitted with a cutter on her bow in case she encountered anti-submarine nets across the straits. No captain could have approached an assignment with greater attention to detail.

At 3.50 a.m. on 19 May Nasmith ordered his boat down to 80 ft (24 m) in order to thread his way through the Turkish minefield at a speed of 3 knots. His crew kept an agonized silence as one mine cable after another brushed along the side of the craft. Eight nautical miles (15 km) further on, as dawn was breaking he surfaced to take a sighting but was immediately forced down by Turkish batteries on the cliffs to either side of the narrow strip of water. Four hours later he was negotiating the final mine barrier. By the early afternoon he was in the open waters of the Marmara but was unable to surface and allow fresh air to circulate until nightfall because of the risk of being spotted. He ordered the submarine down to rest on the sea-bottom.

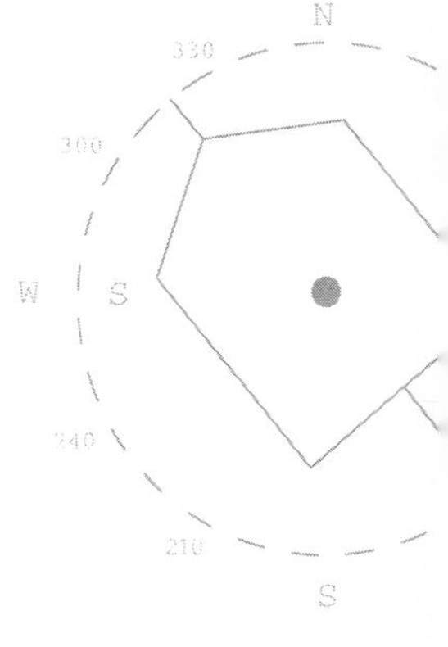

The craft had been submerged for more than twelve hours. The air inside the cramped hull was foul and the grey-faced crew, suffering from oxygen starvation, lay motionless and without speaking inside the iron shell which by then was dripping with condensation. Nasmith eventually ordered the boat to surface at 9 p.m. Seven hours later, as dawn was breaking and with the submarine's batteries recharged, he gave the order to dive and took the E-11 on its first patrol in Turkish waters.

Within a short time of reaching the Marmara, Nasmith identified a phenomenon that might have had disastrous implications for some of the Allied submarines that had tried to get there before him. The straits, and the Marmara itself, were fed by numerous freshwater rivers. Where these outlets joined the sea, patches of less dense fresh water settled on top of the heavier, and therefore denser, salt water of the sea. Nasmith speculated that some of the submarines lost while operating in the Marmara – trimmed for salt water and unable to detect changes in the density of the water outside – would have plunged suddenly and terrifyingly through many feet of water until they hit the dense salt water again. This fall would have destabilized the trim of the boat and might have caused complete loss of control.

Nasmith decided to use the phenomenon to the advantage of the E-11. He knew there were few places in the Marmara where his craft could remain, undetected, to allow the crew essential time for rest. The sea was too small for a submarine to stop, unseen, on the surface and generally too deep for it to rest on the bottom. Nasmith calculated how much water was needed in the trimming and compensating tanks for the E-11 to lie suspended where the fresh water settled on to the deeper salt water. There his crew could relax and rest out of danger and wake up refreshed to continue their patrol. Using ingenious techniques like these Nasmith managed to extend his first patrol to twenty-two days and developed specialized tactics to ensure success.

On the first day of its patrol the E-11 moved freely around the Marmara searching for targets and a safe place to stop to recharge her batteries. However, clear skies and the proximity of the coast on either side made her an easy target on the surface – until one of the crew spotted a Turkish dhow. The E-11 hailed her with a megaphone and drew up alongside. The crew then lashed the submarine to the dhow so that it could remain on the surface during daylight hours protected from enemy look-outs by the sails of the little boat.

Nasmith and his men soon developed a routine which gave them considerable success against dhows carrying supplies and food for the Turkish army and enabled them to replenish their food and fuel supplies and lengthen their stay. The crew would capture a dhow and seize its supplies of fresh water or food such as vegetables or chickens to supplement their own – first taking care to send the petrified crews back to the shore before scuttling their boats. Nasmith sank many Turkish dhows, but adhered rigidly to the rules of restricted warfare. He even unloaded the cargo of one small boat carrying chocolates before sinking her and kept the sweets on the E-11 so that he could give boxes of chocolates to frightened women passengers of other Turkish boats he encountered and sank. However, Nasmith was frustrated by his limited ammunition. He was already sleeping on the metal floor of the submarine with the crew so that he could carry the maximum load, but even so the E-11 had a potential of only twelve strikes. Nasmith therefore set all his torpedoes to float and not – as was normal – to sink so that he could retrieve any torpedo that missed its target. But this operation was neither easy nor safe. The torpedoes were 18 ft (5.5 m) long and taking them to pieces in the water in order to get them back on the submarine was like trying to defuse a live bomb while swimming. One of Nasmith's crew on that first patrol remembered their amazement: 'We were getting short of torpedoes at the time, we couldn't afford really to waste one and the captain said he was going to get it back on board. He dived overboard with a pistol spanner in his teeth, he unscrewed the pistol which is fitted into the warhead, swam back with it, handed it aboard. There were four or five of us on stern waiting for this torpedo and then he went into the water again and we trimmed the boat so that the stern tube was clear of water, he guided this

The crew of the *E-11* with Lieutenant-Commander Martin Nasmith (the central figure on the conning tower), after returning from their first patrol in the Sea of Marmara.

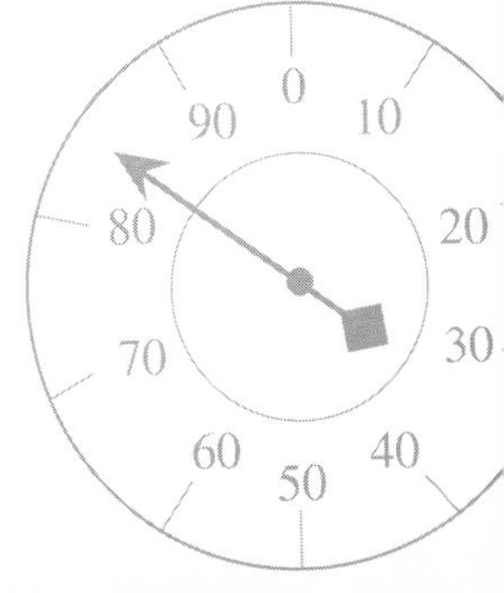

torpedo into the stern tube, warhead first and somebody went from inside the tube and pulled it straight through the tube, through the engine-room, fore end into the foremost tube and we fired the torpedo again.'

On that first trip Nasmith penetrated the harbour of Constantinople where he took the first-ever photograph through a periscope. While there he sighted a prize ship. He aimed and fired but, in his own words, the torpedo went astray and 'travelled twice around the main harbour and up under the Galata bridge, turned hard left and sank a ship alongside the wall'. Not surprisingly, the petrified inhabitants shut up their shops for the day and rushed home.

As his patrol came to an end Nasmith began to think about how he would get back into the Mediterranean. He was concerned by news of a steel net fixed across the entrance to the narrows. He was to write later: 'After the first trip of 22 days inside the Marmara we heard that the Turks were placing a row of large buoys across the straits of Nagara Point which was obviously supporting a net.' The net was made of steel wire and 10 ft (3 m) mesh and Nasmith decided that the only way to get past it was by ramming the wires and eventually diving underneath them. The plan succeeded and, after three weeks on patrol, the E-11 was back in safe waters. Nasmith was promoted to commander. He was awarded the Victoria Cross and many of his crew received decorations. The honours reflected the bravery of all concerned in conditions of continual danger, but the crew did not have long to enjoy them in safety. By August they were heading back towards the Marmara for their second patrol.

Their second and third patrols lasted even longer than the first one and the E-11 – and the other submarines that reached the Marmara – carried a 12-pounder gun that allowed them to greatly increase their striking power.

Nasmith sank a total of some ninety-seven enemy vessels and so frightened the Turks that their fleet was ordered to remain in harbour in Constantinople, leaving few important ships on the open sea for him to attack.

Despite the successes of Allied submarine commanders in the Sea of Marmara, the Gallipoli campaign was an absolute failure. In the meantime, the U-boats had been winning a war of attrition against Allied merchant shipping in the Mediterranean.

Autumn: MARCH 1916 – JANUARY 1917

Lothar von Arnauld de la Perière, the captain of the U35 in the Mediterranean, became the most successful U-boat commander of the First World War. His victories began in February 1916 when he attacked and sank the French liner *Provence* and the British sloop *Primula* on his first voyage as captain. On a three-week patrol which began on 9 June he sank seven steamers and, like his British

Kapitänleutnant Lothar von Arnauld de la Perière, second from left wearing his award for gallantry, and his brother officers on the deck of the *U35* in 1916.

counterparts, used ingenious methods to improve his proficiency. He took on board an 88 mm gun and between July and September 1916 destroyed 90 150 tons of Allied shipping almost entirely by gunfire.

Within ten months of taking command of his own U-boat in January 1916 de la Perière had become the leading U-boat commander both in the number and tonnage of ships sunk. During that time he had already been awarded the *Pour le Mérite*, one of Germany's highest war decorations. On 4 October he sank the French troopship *Gallia* south of Sardinia with 600 French and Serbian troops aboard. He later was to describe it as 'a frightful affair ... the sea became a terrible litter of overturned, overcrowded and swamped lifeboats and struggling men.'

The incredible successes of the U-boat war against Allied supply routes in the Mediterranean gave Germans in general, and the men in the U-boat service in particular, great encouragement. But it was obvious to everyone that the war against supply routes could not be fought only on the fringes. The time would come when it would have to be brought centre stage, to British waters, and that would mean diplomatic tension and, possibly, military confrontation between Germany and the United States. In the meantime, as submariners of both sides went about their arid, frustrating and nerve-racking patrols in the North Sea, reports of the successes of their colleagues in the sunny Mediterranean got back

FELIX
SCHWORMSTÄDT
1917

to them at regular intervals. It was an entirely different war in the Mediterranean, fought in different circumstances and against different opponents, and must have offered to the North Sea crews an exotic and continuing image of success and glamour.

Ritter Karl Siegfried von Georg would have heard of the exploits of de la Perière long before the night he finally decided to bluff his way into trouble in the North Sea. The English fishing fleet was scattered all around him as he surfaced, each skipper evidently concentrating on the progress of the trawl. There were twenty-one boats in all and von Georg knew that he had not been seen. His next move would be crucial because the new rules of engagement were very clear. 'There we were in the middle of the fishing fleet and quite unsuspected. What good did it do? I couldn't sink one unless they chose to let me,' he recalled. 'My orders were to make provision for the safety of crews and the moment I gave any warning, my prospective victim could go scurrying away in the darkness.'

It had been far simpler a year earlier, when Germany had declared the waters around Great Britain and Ireland to be a war zone in which all merchant ships would be sunk with no guarantee for the safety of either passengers or crew. That statement of policy went directly against traditional practice, whereby a belligerent warship ascertained the identity of an enemy merchant ship before destroying it and also made provision for the safety of people on board. But that was before the belligerent attacker was a submarine, and before the British began arming merchant ships for their own anti-submarine defence. By the middle of 1916 any submarine that forced a merchantman to stop so that it could be examined before being sunk risked destruction itself.

From February until the end of August that year U-boat commanders had been free to attack shipping in the declared war zone and, although they had begun slowly, in that final month they sank forty-two British merchant ships representing a total of 135 000 tones. It had been a worrying few months for the British until the *U24* sank the British liner *Arabic* off the coast of Ireland on 19 August with the loss of some forty lives including three Americans. Protests from Washington led Germany to abandon its campaign of unrestricted warfare. From the end of August 1915, U-boat commanders were ordered not to sink passenger steamers without warning and without ensuring the safety of passengers and crew.

PREVIOUS PAGES A painting by the German artist Felix Schwormstädt of a U-boat about to rescue survivors of a torpedoed steamer in the Western Approaches in 1917.

That decision virtually put an end to sinkings for the rest of the year. However, the commanders and their senior officers knew that unrestricted warfare by the U-boats was probably the only way of ensuring a German victory.

In January 1916, Admiral Henning von Holtzendorff, Chief of Naval Staff, argued: 'If after the winter season, that is to say under suitable weather conditions, the economic war by submarines be begun again with every means available and without restrictions which from the outset must cripple its effectiveness, a definite prospect may be held out that, judged by previous experience, British resistance will be broken in six months at the outside.'

The result of arguments like these was a decision to introduce a restricted campaign from 1 March 1916, during which enemy merchant ships found in the war zone were to be destroyed without warning by German U-boats; ships voyaging outside the war zone could be destroyed only if armed; and enemy passenger steamers were to be allowed free passage even if they were inside the war zone. The new policy worked well until the *UB29* sank the French steamer *Sussex* on 24 March killing many passengers including some Americans. The United States' Government protested once again and, once again, Germany climbed down and ordered submarine commanders to revert to stopping vessels, checking their papers and ensuring the safety of passengers and crew of all ships before sinking them.

It had been very frustrating for commanders like von Georg. There had been precious few targets in the North Sea in those first six months of 1916. One of his fellow commanders, the ace Otto Steinbrinck, was to claim that he could have sunk a further forty ships had he been freed from restrictions during this period of the war.

In the North Sea British minefields had become more extensive and anti-submarine nets studded with explosives stretched across some of the channels most often used by the U-boat captains. The German commanders had become frustrated. They felt muzzled and were aware that at some time they would be forced to resort to subterfuge if they were to sink any vessels – even fishing boats. Von Georg had suddenly found himself sitting, unseen, in the middle of a whole fleet of trawlers – and he did not view them as innocent artisans: 'Now a trawler is not an important craft, you would say, but really they were an important adjunct to Britain's sea power. The king's navy relied extensively on Britain's huge fleet of fishing boats. They did all sorts of invaluable drudgery. When they were not fishing they laid mines and swept mines and laid nets to catch the U-boats. They acted as anti-submarine craft, often heavily armed with guns and depth bombs. Sometimes they took the part of Q-ships, trusting to their innocent looks to decoy the unwary submarine commander. And so a trawler destroyed was an appreciable deduction from Great Britain's defence against the U-boats.'

Von Georg regarded the business of sinking merchant ships as disagreeable, but he remembered that night of ship-sinking in the North Sea as having elements of humour that made it exceptional. Earlier on that particular patrol he had taken a Norwegian crew on board before sinking their vessel. He decided to send its captain across to the nearest fishing boat to tell its skipper to abandon his boat because it was about to be torpedoed.

Von Georg knew that his demand was totally illegal under the new laws of engagement. 'It was all bluff,' he remembered. 'If the trawler skipper had refused to obey there was nothing I could have done. For a time nothing happened. I began to think my emissary had used his head and was making off in the boat to which I had sent him. No. Apparently neither the Norwegian nor the English head was working that night. Soon came the sound of many oars splashing. My Norwegian returned and with him the skipper and crew of the trawler. They drew up alongside the U-boat. The mere word "submarine" had brought cold chills of apprehension and evoked perfect obedience. The skipper of the trawler had not even attempted to warn the other fishing boats.'

Von Georg decided to extend the bluff to all the fishing boats. For several hours the splashing of oars resounded on all sides in the darkness. Scores of crowded lifeboats gathered around the black form of the submarine. He put all the crews aboard one of the trawlers and sank the others with gunfire. Then he chased and caught a small Belgian steamer, put the fishermen on board her and sank the remaining trawler.

Effectively, however, the new rules of engagement on the German side and the fact that the majority of British submarines remained in home waters defending Britain against a possible German barricade of the east coast of England, led to very few confrontations between submarines and enemy ships throughout 1916. Nor, apart from the celebrated but inconclusive encounter between the British Grand Fleet and the German High Seas Fleet off Jutland at the end of May 1916, was there any extensive naval activity. A few submarines on both sides blew themselves up on mines or were torpedoed by the enemy. But in northern Europe the months of 1916 were a time of contemplation for submariners and strategists alike.

Like von Georg, Second Officer Ernst Hashagen would have been aware of the exploits of de la Perière the night his boat, the *U22*, was attacked by a British cruiser off Belfast in the spring of 1916. 'There was no chance of a torpedo shot,' he remembered. 'In fact we were the hunted instead of the hunters. She spied us, opened fire and rushed at us to ram us. I lost no time in giving orders to trim the boat down to 15 metres. We dived quickly and got under water all right but something went wrong with the depth rudder. The boat seemed to have gone crazy. She tilted up and down like a rocking horse, sinking now by the head and then by the stern – but always sinking. Down we went to 30 metres, 46 metres, 60 metres. If we went much deeper the terrific pressure would increase and we should be crushed. The only way to rise was to blow the tanks, but that would have popped us out of the water right under the nose of the cruiser up there. Everything else lost its importance in the presence of one particular sound – coughing. I caught the acrid smell of chlorine gas and everybody was coughing, spluttering, choking. My throat and lungs burned with an intolerable torment. The fearful pressure was forcing sea-water through our seams and it was getting into the sulphuric acid of the batteries. Sea-water plus sulphuric acid – any high school student will tell you the answer is chlorine. If we stayed submerged we should quickly be strangled by that infernal vapour. I don't think there is anything that will strike such fear into a submarine man as the thought of being trapped in the iron hull while choking gas seeps from the batteries bit by bit. No death could be more agonizing. It was the old devilish peril of the craft that navigates the undersea, a common cause of ghastly disaster in the early days of submarines. The captain ordered the submarine to the surface. There was no hesitation. No thought of the cruiser up there. Anything for a breath of pure unpoisoned air. Better to be shot to pieces and drown in a quiet way than this death by choking torment. The boat shot to the surface. The cruiser was looming in the mist. Never mind, the hatches were thrown open. Sweet, cool air blows in. We fill our lungs till they almost burst. The cruiser is still there. It stays there. It had not seen us. The fog is dense and blinding and we lie so close to the water that we are invisible. The U22 slinks away through the mist.'

Nikolaus Jaud was the engine-room artificer on the U24, the submarine which had sunk the *Arabic* twelve months before. He too would have known of de la Perière's achievements, but they would have been the last thing on his mind when the submarine's port engine failed while on patrol off the Orkneys. For three days it had cruised without even sighting an enemy ship and he had now been informed that a small-end bearing had cracked. The crew realized that there were few more dangerous stretches of water in which to be stranded with only half their surface speed available if a British warship sighted them. Jaud knew they had no replacement bearings and that the only way to fix the engine was to cast a new one right there on the surface of the North Sea. But with no appropriate instruments or spare metal on board he was going to have to improvise drastically. In the oil-thick atmosphere of the engine-room he made an exact copy of the bearing on a vice and built on to this a mould of bread and asbestos padding to shape the molten metal. He chipped shards of metal from

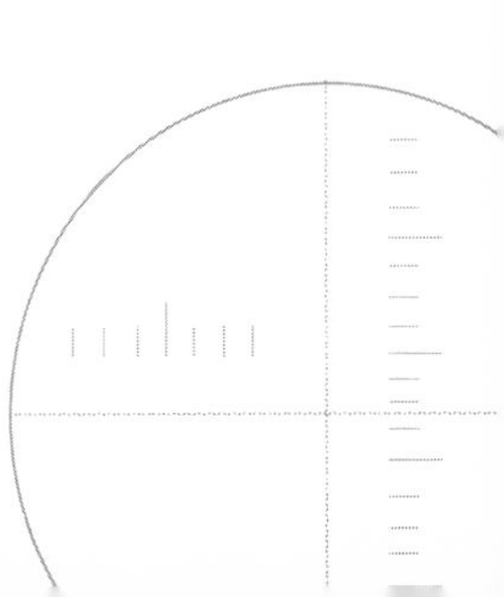

a spare main bearing and then, using the compressed air tanks as makeshift bellows and a floor plate as an anvil, he set about casting the bearing. As the molten metal dripped into the mould his mates jostled to watch, knowing their survival depended on his skill.

As the submarine sat on the surface of the North Sea under cover of darkness, Jaud filed the new cast down to a thickness of one-tenth of a millimetre while the engine was dismantled and the small end of the piston was prepared to receive the replacement bearing. More than twenty-four hours after hearing his captain curse his run of bad luck when first informed of the fault, Jaud reported the port engine back in working order.

A few hours later, with the *U24* moving at full speed towards the dangerous waters off the west coast of England, a second bearing cracked. This fault and the previous one were immediately put down to faulty workmanship in the dockyard at Wilhelmshaven, but for Jaud it only meant one more day of toil in the hot, steamy oil-room where he had first operated his makeshift foundry. Five hours after the *U24* had got under way again he was woken from his sleep to repair a third bearing. By then he was getting used to the work. The third repair took only twelve hours. The next time he was woken it was to learn that his work at sea had allowed the submarine to achieve full speed and catch and sink a large sailing-ship with a cargo of corn.

Jaud knew that his commander, Kapitänleutnant Remy, would be pleased with this outcome. After the *U24*'s previous voyage Remy had had to report that he had made only 50 per cent of hits, which had not been favourably received by his flotilla chief, Admiral Florian Geyer. Remy would not have been made very welcome if he had returned from this patrol to report that a cracked bearing had prevented the *U24* from finding and sinking enemy ships. He too, sometimes, must have been thinking of de la Perière because the regular reports of successes from the Mediterranean would have affected, in different ways, all those submariners patrolling the waters around Great Britain in that dangerous but strangely quiet year of 1916.

Winter: FEBRUARY 1917 – NOVEMBER 1918

Ahead of the submarine and to starboard, the white sailing-ship had spotted, and was trying to outrun, the *U53* which was already moving at its full surface speed of 15 knots. The race was unequal and after a few shells from the U-boat had crossed the sailing-ship's bows it hove to and its crew took to the lifeboats and made for the nearby island of St Kilda. Kapitänleutnant Hans Rose, captain of the *U53* watched as two British patrol boats from the island came out to defend

The *U53* at full speed on the surface of the North Sea in 1917. Taken by a German pilot, this photo must be one of the earliest aerial shots of a submarine on record.

OVERLEAF The crew of the *U53* watch the sinking of a British vessel it had torpedoed. The scene is one of many captured by artist Claus Bergen who went on patrol with the boat and its captain, Hans Rose, in 1917.

CLAUS BERGEN

the cargo ship. They were already shooting and Rose knew he had to sink the sailing-ship quickly if his submarine was not to be caught on the surface.

Claus Bergen, a civilian artist, later recalled his personal dilemma: 'A direct hit had already set the vessel on fire amidships and from it came bursts of reddish flame and thick clouds of brown and sulphur-yellow smoke. The sea poured in through the side of the ship in such volume that, at a distance of about 45 metres, the magnificent creature sank with all sails set in a very few minutes. Only the imperious necessity of sinking every ship within the blockade zone could master the grief natural to every true seaman at the sight of the destruction of so splendid a vessel. But, in this matter, beauty and poetry were as nothing against our duty to the Fatherland. If we were to starve like rats in a trap, then surely it was our sacred right to cut off the enemy's supplies as well.'

The following day the *U53* damaged and stopped British armed trawlers off the Faroe Islands. Before sinking them, crewmen from the submarine brought back boxes and chests of food. Bergen was amazed: 'On the deck was a medley of boxes and chests of cocoa, coffee and expensive tea, sacks of wonderful American meal, fresh butter and margarine, cordage, unused nets, oilskins, rubber boots that did not fit the crew, fine white English bread, English marmalade, ham and bully-beef, bacon and beans, two bars of good soap, tobacco and various oddments. All these things, which were now completely strange to us, we had removed from a few paltry enemy fishing boats, while in Germany the women and children were starving and dying of empty stomachs or supporting life on vile, injurious, almost inedible food-substitutes. The poor in Germany thought of the old days as they sat over their watery turnips while, in the cabins of these trawlers which we happened to have sunk just at dinner time, were plates piled with, what seemed to us, lavish helpings of good fresh roast meat and potatoes, such as we only saw in dreams.'

Chief Petty Officer Roman Bader felt much the same the day his captain targeted a British freighter in the Irish Sea: 'The enemy ship steamed calm and unsuspecting to her destruction. She did not appear to be a U-boat decoy; she had no innocent-looking deck wheelhouse that concealed a gun, no signal halyards between the masts that so often served as wireless antennae, and she lay much too deep in the water to conceal any unpleasant surprises between decks. She was certainly directly helping to destroy Germany and to carry on a system of war that thrust into the hands of innocent German children a slice of raw onion for their supper. We could not spare that ship however much we might regret it. When I travelled about on leave and so often saw children whose angel-souls shone through their pale starved bodies, or soldiers, themselves but skin and bone, carrying home their last loaf to their wives whose hour was nearly come, I was seized with fury against this inhuman enemy who had cut off Germany's food imports. And what I felt, all my comrades on the sea felt too. It was the enemy's crime that forced sailors like ourselves to sink floating palaces, masterpieces of human ingenuity and workmanship. And so this freighter came within our grasp.'

At the end of January 1917 Kapitänleutnant Rose and other U-boat commanders were ordered once again to undertake unrestricted warfare against all shipping around the British Isles and wherever else they could find enemy vessels. Even when their choice of targets had been restricted they had accounted for more than 300 000 tons of enemy shipping every month between October and December 1916. Most of that tonnage, it was true, had been accounted for in the Mediterranean, but enough had been sunk in British waters for Admiral Jellicoe to have become alarmed for Britain's survival in the face of the U-boat menace.

From 4 February 1917 the target for the U-boats was to be more than 600 000 tons every month with the grain ships from North America marked out as special targets. German strategists had argued that such a rate of destruction would account for more than a third of Great Britain's total available shipping tonnage within five months and would have such an effect on food supplies that Germany's principal enemy would be forced to admit defeat, especially since neutral countries caught up in the carnage would be reluctant to risk being destroyed in the U-boat blockade. In addition, the newer, bigger U-boats that were now coming into service allowed Germany to draw the outer ring of the U-boat cordon way out in the Atlantic away from the units of British coastal defence.

From early 1917 a ring of forty U-boats patrolled all approaches to Great Britain on a line that went from the Dutch coast around the top of the Faroe Islands and down to the northern tip of Spain. Britain's plight was stark. If she could not import enough food her people would starve. The U-boat winter had arrived and there was to be no respite.

On 18 March three United States merchant ships were sunk by German submarines and the Americans finally ran out of patience. It took two days for President Woodrow Wilson to decide that the United States had to declare war on Germany and a further two weeks for Congress to agree. By April, when America was in its first month of war, U-boat commanders were sinking an average of more than 800 tons of shipping every day – a combined total for the first months of 1918 of more than 860 000 tons of Allied merchant ships – mainly around the British Isles but also in the Mediterranean where de la Perière and some of his colleagues were still at large.

Jellicoe's greatest fears seemed about to come true. In April, Great Britain had only enough food in the country to last for six weeks. Without some new way of combating the U-boats she would lose the war.

Then in June and July 1917, several U-boat commanders came across an entirely new phenomenon. For weeks on patrol off the British Isles they would see no ships at all and then suddenly a whole mass of vessels would appear on the horizon. Liners and freighters would move first this way and then that as they followed a general course towards Britain, guarded by large numbers of destroyers which circled the perimeter of the fleet. It soon became apparent that the British had inaugurated a new system of bringing food and supplies into the country – a system that prevented lone U-boats from attacking any one of the convoy of ships without taking the chance of being sunk by the destroyers.

The convoy was the result of British and American thinking. It was not a particularly new idea; the British had used it successfully on and off for centuries until the end of the nineteenth century and, since the beginning of the war, they had organized convoys to ferry troops across the English Channel to French and Belgian ports. It was simply that no one had thought of applying it more widely until the summer of 1917 when a huge number of destroyers and submarine chasers were ordered to Ireland from America. By July 1917 thirty-five United States destroyers were stationed at Queenstown waiting to supplement British destroyers and ensure an effective transatlantic convoy system.

But even the convoy system was vulnerable against the instinctive hunter and, after more than two years in the Mediterranean, Otto Hersing in the *U21* was once again operating in the North Sea and the Atlantic. He saw eight convoys during his patrols; he attacked eight times and sank at least one ship on each occasion. In August, 50 miles (80 km) off the south-west tip of Ireland, he spotted a convoy of fifteen steamers in three parallel lines. They were following a zigzag course and were shepherded by twenty-four destroyers. The convoy should have been safe but Hersing held a course between two of the weaving destroyers and came up to periscope height in the half-mile of clear water between the warships and the convoy. He lined up two of the steamers, fired his first and second torpedoes, ordered the *U21* down to 130 ft (40 m) and waited for almost a minute before he heard the two explosions.

He knew the destroyers would pick up on his position from the beginning of the torpedo tracks in the water. 'Every square metre of water was being literally peppered with depth bombs,' he remembered. 'They were exploding on every side of us, over our heads, and even below. The destroyers were timing them for three different depths – 10 metres, 25 metres and 50 metres. They were letting us have them at the rate of one every ten seconds. One detonated right beside us. The boat shivered from the impact and the lights went out. Goodbye *U21* I thought. The lights flashed on again, but the rain of depth charges still continued. We were zigzagging now, more crazily than the steamers above us. But turn where we

Images of U-boat crews as determined, merciless destroyers of the enemy boosted German morale during the last two years of the war when the nation starved because of the effectiveness of the British blockade.

would, we could not get away. The sound of propellers followed us wherever we went, and the bombs continued their infernal explosions. The *U21* shivered with each detonation and so did we. No doubt the destroyers were tracing us by a track of leaking oil from our tanks or with their hydrophones or both. Exactly five hours went by before the hum of that plague of propellers above us died away. For five hours we had been pestered by those blasted depth bombs. How we managed to dodge them all is a mystery.'

Hersing was not the kind of commander to refuse the chance of further attacks because of such an experience. He decided that there had to be another way of reaching the core of the convoy and getting away safely. Instead of trying to put as much distance as possible between himself and his pursuers, he followed the instinct which had taken him to safety after torpedoing HMS *Triumph* two years before and began to dive under the ships he attacked and stay submerged under the very heart of the convoy while the destroyers tried to find him on its fringes. He knew that they could not have attacked him even if they had known where he was. 'A depth bomb thrown there would have done as much, and perhaps vastly more, damage to their own ships as to us.'

But despite such examples of bravery and determination, the U-boats found themselves fighting a losing battle within six months of the introduction of the convoy system. They were still sinking many ships, but from the summer of 1917 the people of Britain were never in serious danger of starvation. And although new submarines were launched at a greater rate than at any previous stage of the war, U-boat losses throughout 1917 nearly tripled. The British were very well aware that the situation had finally swung in their favour.

By the middle of the following year the Straits of Dover were virtually closed to U-boats and the passage between the Orkneys and the Norwegian coast was becoming increasingly difficult. When the U-boats did manage to get into the western approaches many were sunk, either by mines or by British and American warships which were using increasingly sophisticated submarine detection systems before attacking with depth charges. In 1918 these methods also accounted for the destruction of ten U-boats in the Mediterranean, an area where German submarines had previously operated virtually unopposed.

In August 1918 the British claimed that at least 150 U-boats had been sunk during the war. The Germans quibbled but the figure was too close for outright denial. Two months later, at the beginning of October, with the war going badly for the Germans on all fronts, the supreme commander of the German army, General Erich von Ludendorff, advised the Kaiser to ask for an armistice. President Woodrow Wilson stated that no armistice would be agreed unless Germany renounced its policy of sinking passenger ships. On

20 October 1918 Germany announced that all attacks of this kind would cease.

Even then, Admiral Reinhard Scheer, Chief of Staff of the High Seas Fleet, had no intention of allowing the British Grand Fleet to avoid one final confrontation. He ordered the seventeen U-boats still on patrol to take up positions in the North Sea ready to confront the British fleet when it came out of port. But the fleet had been based at Rosyth since April and only a cruiser squadron and some destroyers were available as targets. On 31 October it was all over. The majority of the U-boats still at sea were recalled. The first U-boat war had ended.

On the morning of 19 November 1918 twenty U-boats in a double line moved out of Heligoland harbour and for the rest of that day and the following night cruised at a speed of 8 knots on a direct course towards the coast of England. Before dawn the following day, while it was still pitch dark, the submarines were overtaken by battle cruisers, cruisers and destroyers of the German High Seas Fleet moving to their final anchorage in Scapa Flow.

Leading-Seaman Otto Wehner could not take his eyes off 'these proud, undefeated veterans of the sea. As they gradually disappeared northwards, tears came to our eyes for we were sure we should never see them any more. We had lived through some stern and dreadful hours in the life-story of humanity and in that moment we seemed to endure them once again.'

Some time after eight o'clock on the morning of 20 November the submarine convoy was joined by seventeen English destroyers and a cruiser which led them on towards the coast. 'The cruiser kept on signalling with her searchlights. More and more English ships came out to meet us as we drew in to the coast which, though still ten miles away, was quite visible in spite of the haze; Harwich was clearly to be seen. We kept on course and ten minutes later cast anchor. English officers and men climbed down on to our deck. Then followed some minutes of silent and almost unendurable strain. Our hearts nearly ceased to beat and we bit our lips in defiance of our shame. No, not shame, for we thought with pride of all our victories and heroic deeds and, with sorrow, of our ruined Fatherland. When the English flag was hoisted we turned our backs on it and looked towards our own land and the future.'

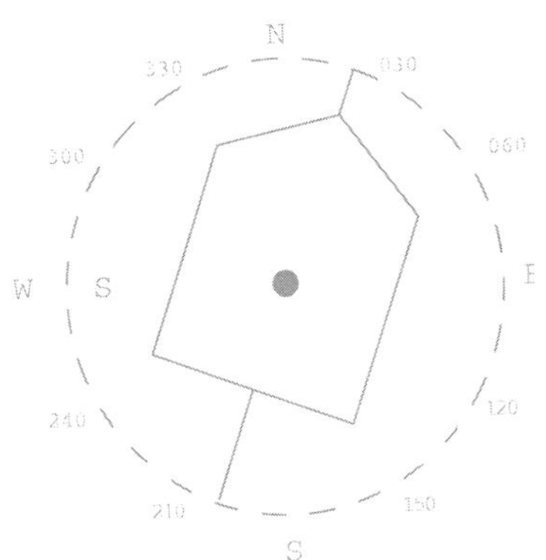

CHAPTER 5

CHALLENGE OF THE DEEP

William Beebe crouched with his mouth and nose wrapped in a handkerchief, his forehead pressed close to the cold window – a transparent chunk of quartz which held back 9 tons of water from his face. 'There came to me at that instant a tremendous wave of emotion, a real appreciation of what was momentarily almost superhuman, cosmic, of the whole situation', he was to remember. 'Our barge slowly rolling high overhead in the blazing sunlight, like the merest chip in the midst of the ocean, the long cobweb of cable leading down through the spectrum to our lonely sphere, where, sealed tight, two conscious human beings sat and peered into the abyssal darkness as we dangled in mid-water, isolated as a lost planet in outermost space. Here, under a pressure which, if loosened, in a fraction of a second would make amorphous tissue of our bodies, breathing our own home-made atmosphere, sending a few comforting words chasing up and down a string of hose – here I was privileged to peer out and actually see the creatures which had evolved in the blackness of blue midnight which, since the ocean was born, had known no following day; here I was privileged to sit and try to crystallize what I observed through inadequate eyes and interpret with a mind wholly unequal to the task.'

William Beebe began his professional career as a naturalist at the Bronx Zoo in New York City but over the years, with the scientist's need to go wider, further, deeper in pursuit of knowledge, his fascination turned from creatures of the land to creatures of the deep ocean.

In the 1920s Beebe had worked off the shores of Maine and Massachusetts. He had also worked in the Caribbean, where he used a copper helmet supplied

'From a coral reef, illumination like moonlight showed waving sea-fans and milling fish.' The magical imagery that first attracted the young William Beebe to the idea of deep-sea exploration.

with compressed air to get close to the natural habitat of the creatures he wanted to study. On one of these dives off Haiti he slid down a rope to a depth of 63 ft (19 m) where his canvas shoes settled into the soft ooze near a coral reef. He made his way to a steep precipice and, balanced on the brink, looked down into the green depths where 'illumination like moonlight showed waving sea-fans and milling fish' far beyond the length of his air-hose. 'It would have been exceedingly unwise to go much further,' he recalled later, 'for the steady force of the weight of water at 10 fathoms had already increased the pressure on eardrums and every portion of my head and body to almost 45 pounds for each square inch. At double the depth I had reached I would probably become insensible and unable to ascend. As I peered down I realized I was looking toward a world of life almost as unknown as that of Mars or Venus.'

These thoughts led Beebe to the possibilities of deep-sea diving. His own immediate idea was to design some kind of strengthened cylinder that would allow its occupant to breathe fresh air pumped down from the surface. On one occasion he discussed this with President Theodore Roosevelt and presented his cylindrical diving craft for discussion. Years later he still had the smudged piece of paper he and the President had used to draw the craft they thought most feasible. Beebe's cylindrical drawing contrasted with a spherical sketch that his companion made on the basis that at depth a sphere would be subjected to equal pressure at all points of its surface. The scientist and the President never returned to the subject but throughout 1927 and 1928 Beebe considered various plans for a cylindrical craft that would be strong enough to sink deep into the ocean.

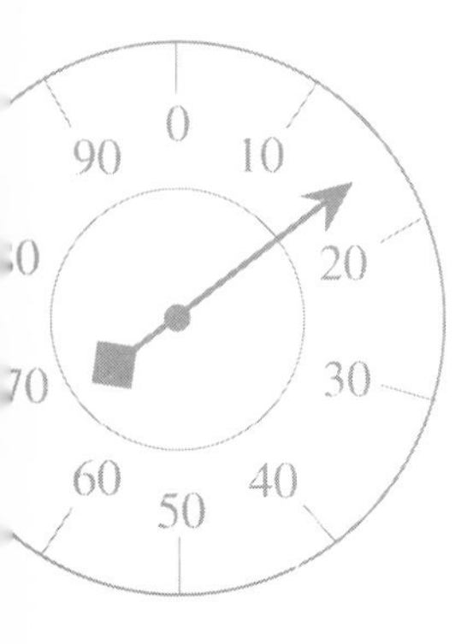

'All of them', he wrote, 'proved impractical. With each 33 feet of depth the pressure of sea-water increases one atmosphere [14.7 lb/6.6 kg] to the square inch, so that at the depth of a half-mile the pressure is over half a ton to each square inch. Any flat surface would be crushed in unless it were impossibly thick or braced by an elaborate system of trusses.'

Beebe speculated on Roosevelt's simple drawing of a spherical craft for many years before he was contacted by a young engineer, Otis Barton, who suggested he help him design such a machine. It was some time before Beebe wrote back to Barton to accept his offer, but it was Barton and a colleague, Captain John H. J. Butler, who were to turn Beebe's dream into reality.

The spherical casing of the craft Beebe eventually called a bathysphere was created from a single casting made by the Watson, Stillman Hydraulic Machinery Company. But the first cast weighed 5 tons and none of the winches available on Bermuda could have handled such a load. The second casting, made of finest steel with walls 1¼ inches (3 cm) thick weighed only 5000 lb (2270 kg) and had an interior space 4½ ft (1.4 m) in diameter.

The sphere was large enough to permit Beebe and Barton to enter and to be sealed up for the descent into, and safe return from, the depths of the ocean. Crucial features which had to be incorporated were windows through which Beebe could observe the marine life at the depths the craft was expected to achieve. The most obvious material to use for them would have been solid glass, but in the completed bathysphere two of the three windows were finished not with glass but with cylinders of fused quartz 8 inches (20 cm) in diameter and 3 inches (7.5 cm) thick which fitted into projections resembling the barrels of very short cannon. Quartz, the strongest transparent substance known at the time, is able to transmit all wavelengths of light. In all, five quartz windows were ground by the General Electric Company but only two survived the fitting and testing stages. The third window was filled with a steel plug.

The entrance to the sphere was a 14 inch (35.5 cm) circular opening which was made watertight for the dives by a steel door weighing 400 lb (180 kg). The all-important question of the air supply was solved by manufacturing it inside the sphere. Oxygen tanks with automatic valves were fitted to its sides with a valve set to release 122 cubic inches (2 litres) of oxygen per minute for the two divers. At this rate one tank would last about three hours. Above each tank was a wire mesh tray. One contained soda lime which absorbed the carbon dioxide and the other held calcium chloride which absorbed the moisture. Palm leaf fans kept the air in circulation. The weight of the craft was taken by a ⅞ inch (85 mm) thick steel cable manufactured so that it would not twist during the ascent or descent.

Shortly after midday on 6 June 1930, 8 miles (13 km) off Nonsuch Island, Bermuda, and 1 mile (1.6 km) above the sea-bed, William Beebe and Otis Barton crawled painfully over the steel bolts at the circumference of the circular entrance-hole and fell into the bathysphere.

There was not even a cushion available to sit on. The discomfort was in keeping with Beebe's mood. He was conscious that as he and his colleague waited to depart on the deepest descent of the ocean man had ever attempted he had not been able to think of any pithy saying that might echo down the ages. 'I had no idea that there was so much room in the inside of a sphere only 4½ feet in diameter and although the longer we were in it the smaller it seemed to get, yet we had room and to spare. At Barton's suggestion I took up my position at the windows, while he hitched himself over to the side of the door, where he could keep watch on the curious instruments. He also put on the earphones.

'At our signal, the 400 pound door was hoisted and clanged into place, sliding snugly over the ten great steel bolts. Then the huge nuts were screwed on. If either of us had had time to be nervous, this would have been an excellent opportunity – carrying out Poe's idea of being sealed up, not all at once, but little by little.'

At 300 ft (90 m) down Barton gave a sudden exclamation. Beebe turned the flashlight on the door and saw a slow trickle of water beneath it. About a pint had collected in the bottom of the sphere. The two men watched the trickle. Both knew that the pressure of the surrounding water would increase with every foot in depth so Beebe gave the signal to descend more quickly so that the metal cover would be forced even harder against the waterproof seal of the surface. The stream did not increase. Two minutes later they were 400 ft (120 m) down. At 700 feet (215 m) the surface team stopped the descent for a while.

'We were the first living men to look out at the strange illumination,' Beebe remembered. 'And it was stranger than any imagination could have conceived. It was of an indefinable translucent blue quite unlike anything I have ever seen in the upper world. I flashed on the searchlight, which seemed the yellowest thing I have ever seen and let it soak into my eyes, yet the moment it was switched off, it was like the long vanished sunlight – it was as though it never had been – and the blueness of the blue, both outside and inside our sphere, seemed to pass materially through the eye into our very beings.

'This is all very unscientific; quite worthy of being jeered at by optician or physicist, but there it was. I was excited by the fishes that I was seeing perhaps more than I have ever been by other organisms, but it was only an intensification of my surface and laboratory interest: I have seen strange fluorescence and ultraviolet illumination in the laboratories of physicists: I recall the weird effects of colour shifting through distant snow crystals on the high Himalayas, and I have been impressed by the eerie illumination, or lack of it, during a full eclipse of the Sun. But this was beyond and outside all or any of these. I think we both experienced a wholly new kind of mental reception of colour impression. I felt I was dealing with something too different to be classified in usual terms.'

After descending another 100 ft (30 m) Beebe inexplicably called a halt to the descent. 'There seemed no reason why we should not go on to a thousand feet; the leak was no worse, our palm leaf fan kept the oxygen circulating so that we had no sense of stuffiness and yet some hunch – some mental warning which I have had at half a dozen critical times in my life – spelled bottom for this trip.'

Five days later, this time with no instinct of danger, the two men were about to pass the 1400 ft (425 m) mark. 'At 10.44 a.m., we were sitting in absolute silence, our faces reflecting a faint bluish sheen. I became conscious of the pulse-throb in my temples and remember that I kept time to it with my fingers on the cold, damp steel of the window ledge,' Beebe recalled. 'I shifted the handkerchief from my face and carefully wiped the glass, and at this moment we felt the sphere check in its course – we felt ourselves press slightly more heavily on the floor and the telephone said "1400 feet". I had the feeling of a few more metres' descent and then we swung quietly at our lowest floor, over a quarter of a mile beneath the surface.'

'As I looked out of my window now I saw a tiny semi-transparent jellyfish throbbing slowly past. I had seen numerous jellyfish during my descent and this one aroused only a mental note that this particular species was found at a greater depth than I expected.

'Barton's voice was droning out something, and when it was repeated I found that he had casually informed me that on every square inch of glass on my window there was a pressure of slightly more than 650 pounds. After this I breathed rather more gently in front of my window and wiped the glass with a softer touch, having in mind the 9 tons of pressure on its outer surface!'

In 1934 Beebe, who had seen the bathysphere examined and wondered at by half a million people in the Hall of Science at the Century of Progress Exposition at Chicago, heard that the National Geographic Society would be glad to sponsor a new dive. But even in the few years since his descent in 1930 engineering science had made huge strides forward. The bathysphere's breathing apparatus of oxygen tanks and chemical trays was condemned as belonging to the Stone Age.

'The old oxygen tanks were scrapped and new ones made to order and fitted with the latest thing in valves – shiny affairs of nickel and glass,' Beebe recorded. 'Even the telephone earphones were replaced. The Bell telephone people said that if I would let them have my old ones for their museum they would furnish sets of the latest models.

'The bathysphere arrived in Bermuda on July fifth. With an impromptu block and tackle we got off the heavy door and took out all the new gear. I prised off the thick, wooden eye plugs and the new quartz lenses gleamed with the sheer transparency of mighty Koh-i-noor diamonds. New steel frames, much stronger than the old ones, held the 3 inch thick masses of quartz as firmly as though they were part of the very steel. In fact, I realized that of the old bathysphere which had carried us down and up so safely nothing remained save the steel skeleton sphere itself. All else had been replaced with more modern, more efficient apparatus.'

The morning of Wednesday, 15 August 1934, saw the beginning of Beebe's thirty-second and last dive. The weather was clear and hot and there was hardly any breeze. The sea was almost dead calm. The depot ship, *Ready*, was 6 miles (9.5 km) south-by-east off Nonsuch Island. By 07.30 hours the oxygen tanks had been mounted in the bathysphere. The humidity-temperature gauge and the barometer followed and by 9.30 a.m. the door bolts had been cleaned and white lead put on the threads to ensure a watertight seal. By 9.50 a.m. Beebe and Barton had crawled inside and just before 10.00 a.m. the nuts were finally beaten home with sledge-hammers.

At exactly 10.05 a.m. the explorers' colleagues, who were once again to monitor their dive and maintain communications, saw the bathysphere swing out, splash into the sea and glide through the ultramarine deep until it passed the vanishing point about 100 ft (30 m) below them. Beebe's comments crackled to the surface.

> *One hundred feet: First Aurelia* [the most common jellyfish off the North American coast].
>
> *Three hundred feet: Pteropod* [the sea butterfly which swims by using a pair of wing-like flaps].
>
> *Six hundred feet: Only grey visible in spectroscope.*
>
> *Seven hundred feet: A mist of copepods* [tiny crustaceans] *and other plankton.*
>
> *Nine hundred and seventy feet: Walls getting very cold.*
>
> *One thousand feet: A shrimp with six pale greenish lights.*
>
> *One thousand and fifty feet: Fish with six lights in a row near front of body.*

Fifty minutes into the dive the scientists on the *Ready* noted that the craft was 2000 ft (610 m) down.

> *Two thousand one hundred feet: Colours of lights are pale blue, pale lemon yellow and pale green. Now two 12-inch fish. One lights up the other then both light up.*
>
> *Two thousand eight hundred feet: Here's a telescoped-eyed fish. It's Argyropelecus and its eyes are very distinct. Marvellous outside lights.*
>
> *Three thousand feet: Siphonophore* [a mollusc], *a big one. Oxygen 1400 lbs. Barometer 76, temperature 77 degrees, humidity 62 per cent.*

When a further 28 ft (8.5 m) of cable had been paid out from the depot ship the crew brought the winch to a halt. Almost half the cable-drum's core was exposed with only twelve or so turns left on the drum. Beebe's assistant on the *Ready* passed him the message that they had reached the lowest point of their dive. The time was exactly 11.19 and 14 seconds.

On all their dives Beebe and Barton had constantly been in awe of the pressure of the water on every inch of their craft. Now, about to be hauled in from a record depth of 3028 ft (920 m), all former statistics must have seemed irrelevant.

'Through the telephone we learned that at this moment we were under a

ABOVE Otis Barton and a thoughtful William Beebe pose with their refurbished craft. Eighty minutes later each window of their craft was resisting 19 tons of pressure as they set a new world record. LEFT Beebe waits for Barton to join him in the bathysphere before the 400 lb (180 kg) watertight door is fitted and the nuts hammered home.

pressure of 1360 pounds to each square inch or well over half a ton. Each of our windows held back over 19 tons of water, while a total of 7016 tons were piled up in all directions upon the bathysphere itself,' the scientist remembered.

The figures had their effect. Beebe's answer to the question that came down from the mother ship more than half a mile above them was that they were ready to be pulled up at once from the black, cold and alien deep where no men had ventured before.

Beebe later described the experience: 'The stranger the situation the more does it seem imperative to use comparisons,' he wrote. 'The eternal one, the one most worthy and which will not pass from mind, the only other place comparable to these marvellous nether regions, must surely be naked space itself, out far beyond atmosphere, between the stars, where sunlight has no grip upon the dust and rubbish of planetary air, where the blackness of space, the shining planets, comets, suns and stars must really be closely akin to the world of life as it appears to the eyes of an awed human being, in the open ocean, one half-mile down.'

At 12.53 p.m., back safely on board the *Ready*, the two men scrambled out of the bathysphere for the last time. Beebe's main concern had never been how far a craft could descend safely under the ocean and be brought back up again. No matter how much he identified with the machine that had kept Barton and himself safe, his expeditions had always been about the marine life he could observe in its natural habitat.

His bravery, and that of Otis Barton, was rewarded by the knowledge that they had been the first. In the years that followed, Barton, working on his own, went even deeper in his own bathysphere while Beebe, unperturbed, went back to his first love – the study of birds in the jungles of South America. But not all his dreams came true. He had believed that his pioneering work would be followed by scores of expeditions in bathyspheres like his that would explore the depths of the ocean and provide scientists with much more knowledge about the vertical distribution of fish than was currently available.

In fact, the man who was to take the first step in Beebe's footsteps would be one of the last to plumb the depths simply to find out what was there. He had closely followed the adventures of Beebe and Barton and identified with the way they made themselves the guinea-pigs in their experiments. Beebe's mantle was about to pass to a superstar.

Outside the German town of Augsburg, only a few weeks after Beebe and Barton had made their first tentative descents in the bathysphere, Professor Auguste Piccard, a Swiss scientist, inflated his balloon for the first time. He had waited several weeks for favourable weather forecasts and on 14 September 1930 his assistants began to pump in the 100000 cu. ft (2830 cu. m) of

August 1932. Piccard first modified his underwater ideas to take a pressurized cabin into the stratosphere to test scientific theories.

hydrogen that he intended would lift himself, his assistant and the beautiful aluminium cabin he had designed 10 miles (16 km) above the Earth and into the stratosphere.

Ironically, the technology Piccard had developed for the balloon and cabin held the secret of how, eighteen years later, he would defy the incredible pressures of the ocean depths freely and without need of a 2½ ton winch cable.

Piccard was a physicist and wanted to test the effects of cosmic rays on the natural gases in the stratosphere, where the vertical displacements of air that cause the condensation of water and the formation of the clouds no longer exist. 'It was to this high region, to be more precise to an altitude of 10 miles, that I wished to ascend to meet the cosmic rays in order to observe them in mass, where their initial properties would not yet have been too modified by collisions with the molecules of our atmosphere,' Piccard recalled. 'A generation had laboured to devise automatic instruments for recording pressure, temperature and humidity. But the measurement of cosmic rays was a delicate operation very different in nature and could not be effected at the time with the necessary precision by these automatic instruments. That is why I decided to ascend myself.'

Piccard calculated that he could inflate the balloon only to one-fifth of its capacity. Then, as it rose the gas would expand in the lower pressure of the rarefied air until it eventually filled the entire volume of the balloon.

That first attempt was postponed even before Piccard entered the chamber, but on 28 May 1931 Piccard waited for the signal from the ground that the balloon had been released from its mooring. Just before 4 p.m. his assistant, looking out of one of the cabin's portholes, mentioned casually that a factory chimney was passing under them. They were on their way.

In less than half an hour they were more than 9 miles (14.5 km) up and well into the stratosphere. 'We had departed before sunrise and we had traversed at high speed those zones where the temperature was between 50°C and 75°C below zero. The walls of the cabin were then very cold and its interior was rapidly covered by a good layer of frost. It was as if we were in a drop of crystal,' Piccard recorded. Twenty-four hours later both men were safely back on Earth. They held the world altitude record but for technical reasons had failed to carry out a single experiment.

Three months later, however, on the night of 17 August 1931, the balloon was inflated once again and before sunrise the next morning Piccard and his colleague were airborne. 'Everything went according to plan,' the aeronaut recorded, 'like a laboratory experiment prepared with minute care. We found out that the particular gamma radiation which, according to a certain hypothesis should have been manifest above in an intense fashion, did not exist.'

For Piccard, who had constructed his balloon and cabin only as a means to a scientific end, all interest in the stratosphere was over. It had never been his true love. Many years before, when he had been a first-year student at the Zurich Polytechnic School, he had read Carl Chun's account of the oceanographic expedition of the *Valdivia*. Chun had described how nets, let down to considerably more than 1000 fathoms (1830 m), had brought back submarine fauna to the deck of the ship: 'When a net was brought up in complete darkness, the oceanographers, leaning over the rails, were struck by the multitude of phosphorescent animals entangled in the net. Certain fish were endowed with veritable headlights. But very quickly these lights grew pale and went out. The fish could no more endure the low pressure and the high temperature of the surface water than we could have endured the enormous weight of the masses of water beneath which they live.'

At that time, as a very young man and before the exploits of Beebe and Barton, Piccard had speculated on the possibility of building a watertight cabin, strong enough to resist the incredible pressure of the ocean depths but furnished with portholes that would allow an observer to admire the world at the bottom of the sea. Such a cabin, the young Piccard reasoned, would be heavier than the water displaced by it and would sink. Therefore it would be necessary to suspend it from an underwater 'balloon' – a large vessel, lighter than water, from which the cabin would hang. He never even considered the idea of a suspension cable linked to the cabin from a mother ship far above on the surface.

The first-year student became an engineer and then a physicist. And when he realized that he had to ascend to 10 miles (16 km) in the cabin of a balloon in order to study cosmic rays he looked to his theoretical underwater observation craft for inspiration. 'The evolution of my thought is clear,' he wrote. 'Far from having come to the idea of a submarine device by transforming the idea of the stratospheric balloon, as everyone thinks, it was, on the contrary, my original conception of a bathyscaphe which gave me the method of exploring the high altitudes. In short it was a submarine which led me to the stratosphere.'

One thing is certain. Piccard's view of the Earth from the cabin of his balloon was more pleasing than the view he and his son enjoyed from the cabin of the bathyscaphe *Trieste* on their first deep dive in 1948 after years of frustration. In between Piccard's ascent to the stratosphere and his descent to the ocean depths, however, other craft, moving secretly under the sea with far deadlier purpose, were to dominate the world's headlines.

Auguste Piccard (right) with King Leopold of the Belgians at an aerostatic exhibition in 1936. Piccard believed a cable far too dangerous. He had always been intrigued by the idea of an underwater balloon that would allow a pressurized craft to float down to the sea-bed.

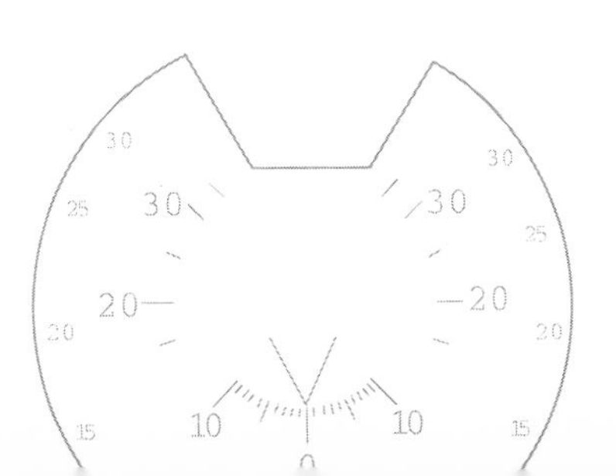

GLOBAL CONFLICT

In the late summer of 1945 Alfie Betts, a Royal Navy instructor in sonar techniques, heard a rumour that the submarine *Rorqual* had returned from the Far East. He went down to the jetty where he saw her, high out of the water and stripped of her batteries, looking very, very pathetic. 'I went up to the ward-room and asked for permission to go aboard because she was guarded by sentries and I wanted to see my old Asdic cabinet. I went aboard, and I had to take a flashlight with me and it was terrible. They had gutted her. And two days later along came a tug and towed the old girl away for razor blades. And I thought to myself: that's not fair.'

At about the same time, Robert Chandler, a radar and sound officer, and his brother officers and crew aboard the United States submarine *Silversides*, returning from the Pacific, had spruced themselves up in their best uniforms for their entrance into New York harbour. *Silversides* was one of the most famous and successful fighting submarines in their country's history. Chandler had not expected a great reception but had thought that there might perhaps be a band to play them in. 'So we go into, I think it was, Staten Island and we put all our flags up and we have our pennants on each side – and there was no one there. There was not one person there to greet us.' A few months later *Silversides* was towed unceremoniously up the Mississippi river to Chicago where she was berthed for use as a training vessel.

Engineering Officer Bernhard Gutschow knew just how both men felt. Nine months earlier he and a skeleton crew had nursed their submarine, *U552*, into a permanent berth at the Wilhelmshaven submarine base. They were asked to make

An artist's impression of a U-boat searching for enemy ships in the North Atlantic in 1943, would have been familiar to Kapitänleutnant Klaus Popp who took over command of *U552* the previous October. Detail of a painting by John Hamilton.

an inventory during which he found pieces of equipment they had never even known about. They found other things, like cigarettes and rum, missing. Finally, they placed the inventory items in a stock room in the shipyard. Gutschow, the last person to leave U552, removed its clock before closing and locking the conning-tower hatch. A few days later he saw the submarine taken into a dock to be scrapped.

For these three men, those private moments were among the most painful of their lives. Against all the odds, *Rorqual*, *Silversides* and *U552*, had fought through years of intense warfare and brought all their men, save one, safely home. There were few submarines that had come so close to destruction on so many different occasions and survived to boast such a record. But when the Second World War had begun six years earlier no one could have foreseen the ferocity with which the submarine war would be fought.

At the end of March, 1940, the British Admiralty decided it was advisable to concentrate considerable forces in the Mediterranean. By the time Italy declared war on 10 June twelve submarines were based in Malta and Alexandria. Ten, including the Rorqual, *had come from China and the East Indies. When Italy entered the war all coasts around the Mediterranean except those of Egypt, Palestine, Cyprus, Malta and Gibraltar were closed to the Royal Navy, which was opposed by five capital ships, twenty-five cruisers, ninety destroyers and ninety submarines of the Italian fleet and 2000 front-line aircraft of the Italian air force. The twelve British submarines were the only means by which the war could be taken to the enemy by attacking his coastline or attempting to sever his supply routes to his armies in North Africa. Submarine operations began on 11 June 1941. Within a fortnight the number of British submarines was down to nine. By the end of July it was eight. It was a difficult time.*

Rorqual: JUNE 1940 – OCTOBER 1941

The first time Leading Torpedo Operator Frank Jordan saw *Rorqual* he was awe-struck. He had never seen such a big submarine. 'It was the sheer size, really, because I'd just come off the H-33 which was a small submarine, only 400 tons with a crew of twenty-two – like being in a cigar. And when I went up to Barrow, there was this great big submarine with a casing about 6 feet high inside which one could walk up and down. The H-33 would have gone on the casing let alone alongside her.'

Rorqual was launched at Vickers Armstrong, Barrow-in-Furness, on 21 July 1936, one of six Porpoise class minelayer submarines. She was nearly 300 ft (90 m) long with a displacement of 2000 tons. Frank Jordan, as fourth Leading Torpedo Operator, was one of the first crew of fifty-nine men. Her two diesel engines produced 3300 horsepower which gave her a top speed of 16 knots on

ABOVE *Rorqual*, her Mediterranean exploits over, in 1944.

LEFT At first awestruck by the sheer size of *Rorqual*, Frank Jordan had long since come to terms with the submarine by the time this picture was taken off Hong Kong in 1938.

During the Second World War torpedo room crews on U-boats like these still ate where they worked.

the surface and 9 knots submerged. She had six torpedo tubes forward. But her main function was to lay mines – fifty of them at any one time.

Lieutenant-Commander Ronald Dewhurst had been in command of *Rorqual* for a year when war came to the Mediterranean. Alfie Betts was on board when he became skipper. 'He relieved Lieutenant-Commander Dennis Sprague – or "Lofty" Sprague as we used to call him,' he remembered. 'We called the new captain "Dizzy" Dewhurst, not because he was dizzy, it was just a nickname. But he was a good skipper and with a skipper like that of course *Rorqual* survived.

'If it hadn't been for *Rorqual*, *Grampus*, *Thames* and *Severn* – they were the big sea-going fleet submarines – especially on their cargo-busting trips from Gibraltar to Malta, I think Malta would have fallen. I do really. *Rorqual* used to carry aviation spirit in her internal tanks and she used to go to Malta with just about enough fuel to get her there. And inboard every nook and cranny was food, boxes of everything you can think of. And when a guy went from aft to forward, or the other way round, he was bent over, treading on boxes of dehydrated cabbage or whatever. Dived all the time of course, surfaced at night, charged the batteries. Daybreak –

down you go again on another little lap and eventually get to Malta. Then you had the minefields that were laid outside Malta by the Italians. We always wondered if there had been new ones put down, or whether the old ones were still there, and so on and so forth. It was a hazardous affair from beginning to end.'

Frank Jordan, so close to Alfie Betts that they were called the *Rorqual* twins, remembers how vulnerable *Rorqual* and her sister ships were: 'When the war started and they shifted them to the Mediterranean, they weren't as good as they thought they were. They were too clumsy, cumbersome, took too long to dive ... well, they were a fair cop for anybody who was searching for them and that was it, that's why we lost so many.'

Frank Jordan had the rank of Leading Torpedo Operator: 'In actual fact we were electricians and one of our jobs was to look after the batteries. We had three on *Rorqual*, each of 112 cells and they weighed more than 8 hundredweights each. And I had to look after these. That was one of the jobs I had to do. I also had to look after the equipment that ran on low power like telephones, the compass and things like that, which was quite useful to learn about anyway.'

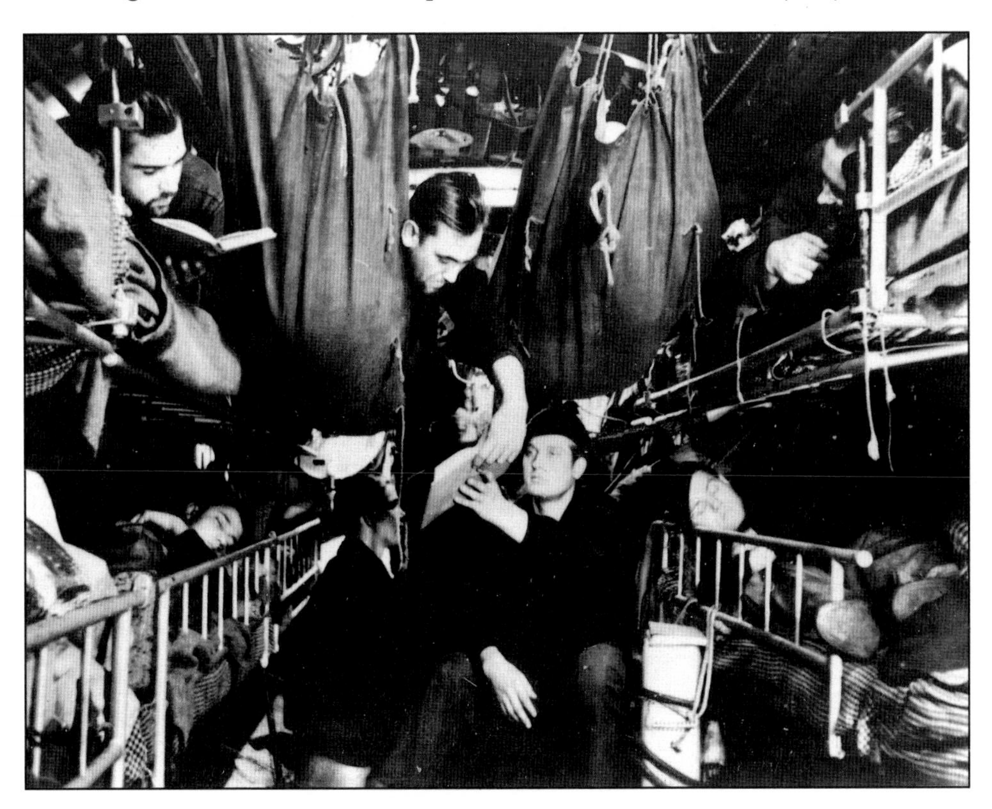

The cramped, crowded sleeping quarters on U-boats would have been familiar to all German submariners throughout the war.

Alfie Betts was a radio operator. 'I was also a telegraphist and an Asdic operator – they call it sonar now, I believe – and therefore I had two jobs. When we were diving I was on the Asdics – our sound detection equipment – as a telegraphist detector and when we were on the surface I was a telegraphist in the wireless office. What can you say about that. I was just a sparker.

'We used to listen out for other ships or anything which was a danger to us. You could, if necessary, transmit and get an echo range off other ships, rocks, sea-bed, whales, all sorts of things. There is an art to listening and not everybody's gifted with it. I know I'm being big-headed – but I was. It's the art of being able to hear things that other people can't hear under water because the speed of sound there is faster than its speed through air. We were trained extensively to listen and detect and, of course, identify, which was the main thing. For instance, when you were submerged an old tramp steamer would come along and the cavitation from his screws was distinctive. You got all these sounds. Diesel was different, turbine was different and so it goes on and on and on. You had to know what type of ship it was just by listening to it. Is it an old tramp? Is it a man-of-war? Is it a diesel-driven job? You get taught all those kinds of things in your training and then you put them into practice when you're in the submarine. And all this is reported instantly to the officer on watch or the captain, whoever's in the control room.

'Oh, the sea-bed causes a heck of a lot of interference and you've just got to get used to those noises. You push them into the background and get the ones you want. Whales? Oh, they're beautiful. Hear them talking to each other, killer whales especially. And dolphins, dolphins are delightful. They've got a noise that sounds like a sonic echo-sounder. Crustaceans, you hear them on the sea-bed, crabs and things like that. Oh yeah. You can hear all that sort of thing. And you learnt to know all these sounds. You had to, for your own safety.'

Frank Jordan remembers that the *Rorqual*'s first trip out was to Brindisi on the eastern coast of Italy. 'We had to mine the harbour and we put two rows of twenty-five mines across the mouth. They couldn't have heard us because they didn't come out, but we just led off and watched through the periscope. We saw two ships hit them and we disappeared off to do a proper submarine patrol. I think we were out for three weeks or less but when we entered the harbour at Alexandria, everybody was saying that there should be some boats alongside us. There was nothing on the starboard side where four others should have been. We were the only boat to come back from that patrol. We heard afterwards that all the others had been sunk. Then three of us went out again on the next patrol. We did a skirmish with a couple of ships which we sunk and we were depth-charged but it was nothing. They used to drop their depth charges yards away.

And we came back in and again we were the only ship to get back. Afterwards no one wanted to go out with *Rorqual* because they thought she was a jinx.'

Rorqual always operated alone and her first priority was to lay her fifty mines across shipping lanes used by enemy merchant vessels and warships. Only after disposing of all her mines did she look for targets she could attack by torpedo or with her deck gun. And only when her ammunition was gone or her supplies were running low did she return from patrol to her base in Alexandria.

Dewhurst took the cumbersome *Rorqual* out on mission after mission in the clear waters of the Mediterranean. Running submerged by day and making every attempt to avoid offensive action until all the mines had been laid, he surfaced only during the hours of darkness. Between 10 June 1940 and 23 May 1941 *Rorqual* carried out ten patrols totalling 156 days at sea during which she laid 450 mines in enemy waters and sank a tanker, two transports, a submarine, an ocean-going tug, a schooner and a caique. In the first ten months of 1941 Lieutenant-Commander Dewhurst was awarded the Distinguished Service Order three times and *Rorqual* was seen as a successful boat.

Dewhurst's last action, shortly before he left *Rorqual*, was to attack and destroy a Greek caique flying a German flag off the island of Lemnos. He ordered *Rorqual*'s gun to fire at the caique. There was a big explosion and a white flag was run up. Four German soldiers on board the caique were killed and the Greek crew abandoned ship before *Rorqual* again fired at the boat to sink it. It was 11 May 1941. Less than three weeks later Lieutenant-Commander L. W. Napier took command of *Rorqual*. At the time there was no reason for anyone to suspect that the incident with the caique might have led to Dewhurst leaving the ship.

One of the first things Lieutenant-Commander Napier did was turn night into day for *Rorqual*'s crew. 'I thought it convenient,' he said, 'to have breakfast in the evening, our midday meal at midnight and supper in the early morning. The advantages of this were that at night, on the surface, when it was more likely that a sudden emergency might arise, most people were awake.' On *Rorqual*, as on all submarines, no cooking or smoking was allowed when the craft was submerged and the change of routine had important benefits for the crew since they could use valuable time on the surface during the hours of darkness. As soon as dawn broke they would have been an easy target for German and Italian warplanes if they had remained there.

But there was something much more disturbing in *Rorqual*'s immediate past about which the crew were still talking and about which the new commander had been briefed by Lieutenant-Commander Dewhurst before he left the boat. 'My predecessor told me of an incident which had occurred a short time before, when he had been involved in sinking a caique in the Aegean which had German soldiers on board. In the course of this action, the German soldiers, or at least some of them, had been killed and the manner in which this occurred had led, subsequently, to some controversy as to whether this had been right. As I understood it, they had stopped a caique on the surface in daylight and had ordered the crew to abandon ship in their boat. They had done so. They were, of course, no great distance from land and the weather was good. Then, to the surprise of members of *Rorqual*'s crew who were on the bridge, a number of German soldiers in uniform appeared out of the caique's hold. Exactly what happened after that, I don't know. But in the end it led to the shooting of at least some, if not all, the German soldiers in a manner which was, perhaps, thought to be a rather cold-blooded killing of the enemy.

'I, of course, knew nothing of it directly, not having been present and, indeed, hadn't even heard of it. But Dewhurst did tell me that this had happened and that it had led to the death of some German soldiers in, perhaps, doubtful circumstances. But he told me that if anything of this kind were to occur in future, or if I had to act in this manner, I would be backed up by the ship's company who were not unduly distressed about what had happened. I was never happy about this incident, and shortly after I'd taken over I said to all my officers – I didn't say anything about it to the ship's company – that I didn't really like what had happened and I very much hoped that no such incident would happen again, or I would be placed in a situation where I felt I had to act in this matter.'

Lennox Napier saw his responsibility in the Mediterranean very clearly. 'Our task was largely to prevent the enemy making the best uses of the advantageous circumstances in which he found himself. It was our job to cut supplies to North Africa where the German armies were active and, by attacking shipping of all sorts around the coast of Greece and the Aegean, to reduce the enemy's potential to supply himself and his scattered forces by sea.'

In the Mediterranean theatre of war, the island of Malta was the key to everything. If it had been lost in the desperate siege years of 1941 and 1942, British prospects of defeating Rommel's army in North Africa would have been bleak. The strategic importance of the island was clear to Lennox Napier, as it was to everyone fighting in the war: 'Malta was on the direct line of communication between Italy and North Africa. It was right on the route along which the Germans had to run all the supplies and reinforcements for their armies in North Africa and therefore it was an extremely good base from which to operate against them.' While Malta could be kept going there was always the possibility that Rommel could be held.

When *Rorqual* left the Mediterranean for a refit in Great Britain in October 1941, the battle for Malta was still raging.

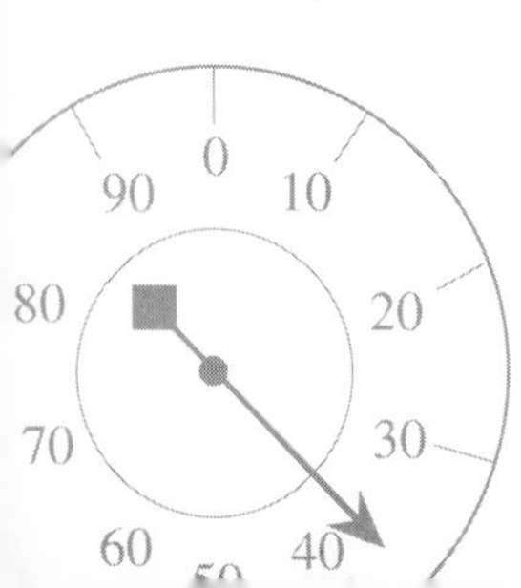

As soon as Great Britain declared war on Germany on 3 September 1939 both countries reverted to the strategies they had used with such success in the First World War. Great Britain declared a siege on Germany by means of a naval blockade. In return Hitler, who charged Britain with conducting not open warfare but 'the mean and brutal starving out ... of the weak and defenceless, not only in Germany but in the whole of Europe', declared a counter-blockade using his U-boats. At the outbreak of war Germany had fifty-seven submarines of which only thirty-nine were fully operational. A large-scale construction programme of more than 300 U-boats in twelve months was put into action but suitable metal was rare and only thirteen boats had joined the fleet by April 1940.

Four months later, eighteen of Germany's best submarines were ordered into the Atlantic to sink as many merchant ships as possible before the British could organize a convoy system to protect the million tons of shipping that had to arrive safely at her ports every week if she was to wage a successful war against Germany. In the first six months of the war the Germans sank an average of 142 453 tons a month. Between June and October 1940 the monthly average leapt to nearly 300 000 tons. The U-boat men called it the 'happy time' because of the high level of sinkings and their negligible losses. The decision to remove convoy escorts to help evacuate British troops from Dunkirk left merchant ships relatively undefended and gave the U-boats a free hand in the Atlantic for some time afterwards.

U552: JANUARY 1941 – DECEMBER 1941

Kapitänleutnant Erich Topp was given command of U552 in December 1940. She was one of the new VII class of U-boats built by the Blohm and Voss yard at Hamburg. Topp had lost his previous submarine, the U57, after returning from his third cruise around the British Isles, when it had been in collision with a Norwegian steamer. Six men had been killed but when Topp was given command of U552 he took the rest of its crew with him. The U552 was a much bigger boat than the U57. The VII class was the pride of the U-boat fleet and was 218 ft (66 m) long and displaced 1000 tons when submerged. Its diesel-electric engines delivered 18 knots on the surface and 8 knots submerged and it was armed with four forward and one aft torpedo tubes and a 4 inch (10 cm) gun.

It was to become known as the 'Red Devil boat' because Topp had had two dancing devils painted in red on the conning tower that rose menacingly above the flat deck. 'They weren't my original idea,' he said. 'They had existed on the U57 and had been put there by its previous commander, Klaus Kort, before I took over command. The two figures represented "extermination" and "life" which was especially important considering the losses of submarines that had already occurred.'

ABOVE The conning tower of the Red Devil boat *U552* in 1941. Erich Topp, three times personally decorated by Hitler for bravery, is silhouetted top left. RIGHT 'Our goal, brutal as it sounds today, was to sink ships; to bring England to its knees.' Erich Topp, in 1941, before he began to suspect the British had broken the German codes.

Topp saw his job in very simple terms: 'Our goal, brutal as it sounds today, was to sink ships; to bring England to its knees. There was no particular political motivation in the navy at the time. When I joined the submarine flotilla in 1937, Germany had just awoken from a long sleep. She had tried to gain political acceptance and our political leadership promised to untie us from the bonds of Versailles. According to the declaration of Versailles, the final reparations for the First World War would have been in 1990 – the year of Germany's reunification. The scale of this is forgotten in many people's minds.

'At sea, daytime and night-time did not exist for us. We were under constant alert, always ready for an alarm. There were days when a commander got no sleep at all. When a convoy was being hunted – and this sometimes took days – he did not even leave the bridge unless it was necessary to dive.

'There were also peaceful times, times when we could rest. For example, when cruising in the middle of the Atlantic or when we were called from one operation to another at a different location. That time travelling was spent conserving energy. One thought about what had happened and about future plans. At the beginning of the war there were even times when we played cards or listened to records if the weather was bad and we were sure of not being attacked. But such times were exceptional.

'The submarine operational leader's mission was to attack convoys in the Atlantic. To do this, one has to know where the convoys are cruising. We were informed of this from time to time by our intelligence people who were very good. They could tell us sometimes where the convoys could be found. We then put together "outposts" to catch them in the high Atlantic. Such a plan consisted of a line of U-boats strung out over fifty or more sea-miles. The leading ship of the convoy leaves a trail of smoke. One U-boat would spot it and then signal the other submarines in order to inform them and get their help in attacking it.

'On 31 October 1941 we attacked and sank a destroyer which was escorting and guarding an English convoy. The destroyer turned out to be the USS *Reuben James*, an American ship. We first launched a torpedo which broke the destroyer apart. Then, when it had sunk about one metre, its depth charges fell on top of it creating an enormous explosion. At first I thought we had sunk a British ship. Then we found out that it wasn't and I began to think. A whole load of thoughts crossed my mind. We had sunk a ship of the United States which was not yet even taking part in the war. I knew the role of submarines had been significant in the First World War and that my actions were not to be the best of influences on the state of the war. All this crossed my mind. I was frightened of the political complications for which, in the end, I might be responsible. I knew, however, that I had acted under international law.'

Japanese successes in the first six months of the war in the Pacific were achieved, maintained and exploited by its immense fleet of military and commercial shipping. The war leaders in Washington recognized that only by destroying those fleets, and particularly the merchant fleet, could Japan be stopped and eventually strangled into submission. The only answer for the United States was the submarine – set free to strike hard and often at the 6 million tons of merchant shipping that supplied the civilian population in Japan and the Japanese army which was rapidly spreading across the South Pacific. There were only twenty-eight submarines in the Philippines to take on the entire Japanese fleet but in January 1942 they started to sink enemy ships off the coasts of Japan. During the first two years of the war in the Far East, United States submarines accounted for 73 per cent of Japanese ships sunk in the Pacific. From the moment they left harbour they were intent on unrestricted warfare.

Silversides: DECEMBER 1941 – JULY 1943

Lieutenant Robert Worthington reported aboard the submarine *Silversides* one week before the Japanese bombed Pearl Harbor on 7 December 1941. He had come from another fleet submarine but it was of an older vintage and he was thrilled to be on board the newest submarine in the United States navy. *Silversides*, commissioned two weeks later, was air-conditioned and fitted out for comfort. A large, fast craft, she displaced 2424 tons when fully submerged and was more than 300 ft (90 m) in length. She could travel at 20.25 knots on the surface and nearly 9 knots when submerged. Like the other new submarines of the Gato class which followed her, she had a complement of eighty officers and men and was armed with ten torpedo tubes and a 3 inch (7.6 cm) gun.

Worthington was responsible for the boat's hull and weapons under its skipper, Lieutenant-Commander C. C. Burlingame. He remembered Creed Burlingame as a very gung-ho, very aggressive officer, but a friendly and easygoing man socially. 'He was all business at sea and very efficient and very effective as a commanding officer. He had skippered an older submarine in the Philippines before he came to *Silversides*.

'We were very eager to get to the main islands of Japan and find some Japanese ships to sink. We had been through four months of fitting out, training, practising and readying ourselves for this operation and we could hardly wait to get there. We all had a hatred of the Japanese at that time because of the treachery of their attack on Pearl Harbor.'

Radio Operator Sam Remington remembered feeling detestation more than hate when the *Silversides* berthed at Pearl Harbor on its way to the South Pacific in April 1942. 'We were able to see, when we pulled into Pearl Harbor, what they had done. The damage; the partially sunk ships turned over on their sides. And it was almost as though they had done it to us; as if we were part of it, although we weren't at the time that it happened. And we wanted to do as much damage as we could to them. And that was our attitude and that was our feeling. As individuals there was no real hate.'

For pharmacist's mate Thomas Moore, who had helped care for the dying and wounded at Pearl Harbor during and after the Japanese attack, anything that was firing at the Japanese was all right by him: 'I wanted to see the Japanese get whipped in any way we could whip them,' he said. 'I never did feel bad about the injuries we gave the Japanese and I've never felt sorry about Truman dropping the atomic bomb, because if any nation ever deserved it, they did. And that's my own gut feeling on it.'

Moore was to have many chances to enjoy his revenge under *Silversides*' first commander. 'Creed Burlingame was a leader of men. He was fabulous. Every man on that ship would have sawn off their arm if they thought it would have helped old Burlingame. But he was just that kind of man. When he said something he meant just what he said. And he intended it to be carried out if it was an order. If you had a problem of any kind, why he'd listen to you. He was just fabulous. He was very successful at sinking lots of shipping. He had excellent discipline among his crew and he didn't do a lot of fiddling around. Whenever he said "Up periscope" and "Make ready one and two and three for'ard" that's what he meant, and you could expect to see something go off the ship pretty quickly because he was very decisive.'

Remington remembers Burlingame equally vividly. 'Well, he was a he-man type, a man's man. Hard-drinking, hard-fisted. Burlingame once made a statement saying that on board *Silversides* ranks and rates were left on the gangplank. You respect a man or you salute a man for what he is, not what he should be. In other words, an officer was an officer because he was a good officer not just because he was an officer. And that's the way we respected our officers. And they, in turn, the same. At sea it was very informal. But in port it was just the other way around. We enjoyed saluting our officers because they were worth it.'

On one of *Silversides*' trials Remington was left on deck as she submerged and was only saved because someone had taped open a telephone connection to the bridge. He thought he was about to drown, but the only man the submarine ever lost was killed one month later, in its very first taste of action, when Creed Burlingame decided to attack a Japanese trawler with *Silversides*' deck gun.

OVERLEAF USS *Silversides*, 'air conditioned and fitted out for comfort', about to be launched in the autumn of 1941. A few weeks later the Japanese attacked Pearl Harbor.

OFFICIAL PHOTOGRAPH
NOT TO BE RELEASED
FOR PUBLICATION
NAVY YARD MARE ISLAND, CALIF
236
U.S.S. SILVERSIDES
A
MARE ISLAND
PRODUCT

23

One of Remington's jobs was to jerk the hot shell-casing from the gun after it had fired so that a new shell could be inserted into the breech. 'Mike Harben was carrying shells from the conning tower to the gun and as he went back to get another one I'd grab the empty shell. I'd kick it overboard – we weren't taking it back, we just kicked them overboard. Suddenly Mike went down in a spray of machine-gun fire. He was standing right next to me and I had a small beard and I had a burn mark right across my cheek. The beard was gone and there was just a little red mark there. And Mike, being a little taller than I was, was hit right under the rim of his helmet by a machine-gun bullet. We grabbed him, we carried him into the conning tower and as we put him through the opening, the bullet fell out of his helmet on to the deck. And afterwards we always felt we were all lucky. He was the only one. He took all the bad luck away from us right there.'

Worthington would agree with Remington about luck: 'At one point I was crouching down behind the thin plating around our bridge area as the patrol boat returned our fire and one machine-gun bullet came half-way through the plating in front of my eyes and stuck there. The point of the bullet was on my side of the plate and its tail end was on the Japanese side.'

On *Silversides'* third patrol Remington was made radio man and from that moment on his battle station was always in the control room. 'It was just one of those things. Why me? I don't know. But I guess it's something I did well. At battle stations I was right up there in the conning tower and as long as we were submerged I was the eyes and ears of *Silversides*. That's the only thing; they had to know what was going on other than what the ship itself was doing. It was my job. To me it was not a responsibility, it was a job you do and you do it well. That's the way it was with all the men aboard. When you qualified you had to be able to stand a watch in every compartment in the ship. You had to do a little bit of everything because if you were trapped in a particular compartment for some reason and couldn't get out you would have to do whatever was necessary to make that compartment function.'

'My job was to listen to the screw beats. It was a distinctive sound – a swish, swish – and you could count the speed of that swish and if you knew the type of ship it was, you could almost tell what its speed was. You could count the number of turns, as we call them. Every swish was a turn and every turn allowed you to compute how fast that ship was going. And if the screw beats were going normal, and all of a sudden speeded up, you knew they were going to make a run. They'd found something that they were going to do. And this is it. No one except the man on the sound gear has any idea of what's going on. He is the one. Everyone else just stands. And you have to convey that information to the skipper who, in turn, conveys it to whoever he wants. And if the ship is

A reconstruction of *Silversides'* sleeping quarters as they might have been during the war. Sleeping conditions on US submarines were relatively luxurious and spacious. There were even sun lamps on some boats.

making a run on you, and you know it's making a run on you, the skipper's going to try and do some evasive tactics – maybe make a right … a port or starboard turn … maybe speed up a hair because when the tracking ship goes to full speed its crew can't hear us. They have to slow down for their sound gear to hear us. So then we can speed up a little and make a turn or whatever the skipper wants to do to manoeuvre and avoid the oncoming ship.'

When it came to seeking out and destroying, rather than hiding and escaping, Creed Burlingame refused to move away from old-fashioned methods. 'Creed was an old-time submariner,' remembered Worthington. 'He was not used to the high tech of the time. He was used to manual computing, calculating enemy track courses and speeds and predicting the proper lead angles for your torpedoes to impact the target. He did not believe in anything as fancy as the newly introduced magnetic exploder for torpedoes. So, instead of firing our torpedoes to go underneath the ship where they would be activated by its remote magnetic field, Burlingame fired his torpedoes shallow, for direct impact, and we were very successful in our first four runs. As I remember, 63 per cent of the torpedoes we fired actually sank ships in our first four patrols.

'And when we did hit, it was always elation at having eliminated part of the Japanese war machine. But you usually had a feeling of something akin to sorrow that you were putting a ship down, because ships were our business. Strangely, you never thought about the people on board the ship who were dying. All you thought about really was the ship that was dying. It was a sad experience for a mariner to destroy a ship.

'And as ships sank close by you could frequently hear their bulkheads collapse. Ships under attack or in wartime conditions had all their watertight doors shut and the compartments segregated as much as possible. And once a ship went down, bulkheads would collapse as the pressure increased and cause loud bangs. It was like a continuous crumpling sound as metals crumpled like sheets of paper.

'We had full confidence in the captain's ability to evade the ships that came after us after the attack. Fortunately, they usually ran out of depth charges or they ran out of daylight. I don't believe the Japanese liked to conduct war at sea in the dark. When the sun went down, they tended to go back to port so we would usually be able to surface by eight or nine o'clock in the evening when we were under attack from Japanese anti-submarine forces.'

Remington remembers feeling sorry for the men who had nothing to do during an attack. 'During a depth charging, or even during an approach, most of them have absolutely nothing to do. All they can do is just sit, and they have time to think. And to me that is the scary portion of being aboard a submarine during an approach or a depth charging. I don't think I could have taken it myself, but I was lucky, I was right in the middle of everything. But I sure felt sorry for them. When a depth charge was dropped you would hear the click before the charge exploded. Then you'd hear the rush of water going through the superstructure which was all open and the ship would rock. And if they dropped it right on top of you the whole ship would shake. It seemed as if it would come up in your face and then it would just continue to shudder for a second or two. Other than that you'd hear nothing. It would be just as quiet as could be because during a depth charging everything that would carry sound is cut off in the ship. So you wouldn't hear anything; no air-conditioning, no fans, nothing. It's very, very quiet.'

The number of U-boats climbed steadily to 109 by March 1941 but because of training needs an average of only six or seven patrolled the Atlantic in search of convoys at any one time. Between April and December 1941 the number of U-boats more than doubled to 250. This expansion allowed the implementation of wolf-pack tactics against the convoys and brought about the full fury of the battle of the Atlantic with up to twenty-seven U-boats at sea at any one time. Groups of them, forced west into the open Atlantic by improved British anti-submarine defences, patrolled in lines across a convoy's probable line of advance.

But although there were some notable successes, the patrol line tactics generally failed to intercept as many convoys as expected. Suspicions grew that radio transmissions from U-boats allowed the enemy to locate them with direction-finding equipment and warn convoys to take avoiding action. Decreased use of the radio did not change the situation and the U-boats were therefore ordered to turn back from the empty central Atlantic into the heavily guarded waters around the United Kingdom in search of targets, which meant that the convoys were even more successful. The monthly tonnage figures of merchant-shipping losses fell from an average of 309 000 between April and June to only 99 000 in July and August. This was because the British, who had already broken the German naval code, captured the U-boat Enigma machine-code settings on 11 May 1941 and had complete mastery of U-boat communications by August.

In almost two months from mid-September only five convoys were sighted and attacked. Soon afterwards, a baffled U-boat Command abandoned its broad patrol lines strategy of convoy interception.

U552: JANUARY 1942 – SEPTEMBER 1942

Erich Topp had a feeling that his wireless transmissions back to U-boat Command headquarters at Lorient were no longer secure and that the codes he and other submarine commanders were using had been broken. 'We reported this but the leadership assured us again and again that the Enigma machine which encoded and decoded our messages had billions of options and that it would be absolutely impossible to crack our codes. Still, my crew and the other boats had a feeling that the other side had the power to break it.'

But Topp was also becoming aware of certain limitations in the radar devices of the destroyers that hunted U552 after its attacks on convoys. 'In one attack on a convoy heading towards the British Isles from Gibraltar we launched all four torpedoes and, as we turned to get away, a destroyer which had spotted us began to come at high speed towards us. It was clear if we were to dive we would become his victim. So I tried to run on the surface to get away from him using the combined power of my diesels and batteries. It was dark, we could only see for 700 metres at the most. The destroyer came closer. I could already see its bridge. Obviously they had us on their radar system. Suddenly they launched depth charges, but to our luck we got away. Later we found out that radar systems could only recognize a target beyond a range of around 500 metres or so. The destroyer obviously lost us on his radar system and thought his attack was successful. Our

LEFT 'What will the new commander be like? Will he be able to get us back safely?' A German submarine leaves for another patrol during 1943.

BELOW A U-boat returns from patrol to join the flotilla at anchorage in Kiel during 1943.

other U-boats waiting there did not get a chance to attack. The convoy security system was very good.'

Topp and other U-boat commanders began to find that as they approached a convoy the defending warships would leave it to intercept them, or aeroplanes would come out of the sky to attack them as they closed on the convoy, forcing them to stay hidden under water. The ability of the enemy to know exactly where the submarines were convinced Topp that the U-boats had been compromised.

As they moved around the Atlantic waiting for targets the crew of U552 was forced to endure extremes of temperature. On one occasion when the men were kitted out for tropical weather, Topp was given orders to go to Newfoundland. This meant cruising from the Gulf Stream to the cold north and while the boat was submerged its crew experienced changes of temperature as sudden as 15°C within one minute. 'Such were the changes in water temperature and also in the boat. We never had heating. Well, we had some, but it had never been in operation because we always had to save battery power. Near Newfoundland the temperature was minus 10°C on the surface. This naturally meant that any water on the submarine immediately changed into ice. So we had to dive every three hours to deal with these weather conditions. It wasn't particularly easy. Some people returned from that cruise with frostbite on their hands and feet. We didn't have much success either because – I don't know really – there was just not so much traffic in that area. It was very minor, we sank two boats there. So we cruised on to the south. We were at the end of our capacity and we did not have any success. It was complete misery. Shortly afterwards on our second American cruise we operated at a place where the main traffic was sailing. We sank ships in a relatively short period of time. At the time the American defences were rather weak. We had seen planes and even blimps above us from time to time. We watched in peace. They didn't harm us because they were just too slow to launch bombs at us. We had shot all our torpedoes and had sunk about 40 000 tons so we cruised on home. That second cruise was a relatively big success.

'At the beginning of the war we operated in the Atlantic close to the European continent. That changed as the resistance in that area got stronger. So we moved on – further into the Atlantic, into the South Atlantic and even to the Indian Ocean. There it was still possible to get support from above water. Ships supplied us with oil, torpedoes and food. That wasn't possible in the Atlantic any more. That's why it was decided, relatively early on, to build up an underwater supply system – the so called "milk cows". These supply submarines were around 2000 tons and carried all the necessities like torpedoes, fuel, oil and food supplies. We met with them at pre-planned positions where we hoped none of the

By August 1942, more than a hundred U-boats were available to patrol the Atlantic looking for targets. While in harbour submariners visited other boats to talk about the war.

opposition was operating. But the enemy got to know our rendezvous positions and they were all sunk.

'I began to realize the sudden change while I was still on the U552. Aeroplanes began attacking us suddenly. Planes that we didn't see, just appeared through the clouds having already located our position. As the statistics began to show a decreasing number of ships being sunk and an increasing number of U-boats being sunk, the trend of the figures made me realize that we would lose the war.'

Hans Babel had joined the U552 in the spring of 1942 when one of Topp's crew opted out. As the boatswain, he remembered having an unbelievably wide workload and responsibility for all the equipment like cooking pots, dishes, rain uniforms, helmets and outfits to be used on duty in the conning tower. He worked closely with Topp. 'We had great respect for the old man and the whole crew looked up to Commander Topp, especially since he was well known as a hero at the time. He had received high honour and distinctions. When we heard there was to be a change in commanders then, you can imagine, we were very anxious how things were going to work out with a new commander on U552.'

Throughout the first half of 1942 the situation on Malta deteriorated rapidly. The German air force bombed it continuously from southern Italy. In the first four months of the year the island was on alert twelve hours out of twenty-four. Many British submarines, needed for offensive operations against Rommel's supply lines, were destroyed in these attacks and the flotilla based at Malta was forced to leave for Alexandria until things cooled down. By the time Rorqual *returned to the Mediterranean after her refit, British submarines were disrupting Rommel's oil and petrol supply lines as the Eighth Army in North Africa readied itself to attack at El Alamein.*

Rorqual: AUGUST 1942 – OCTOBER 1943

Tom Johnson worked in the engine-room but when Lieutenant-Commander Napier wanted to attack it was his duty, as the artificer, to tend to the periscope and take it up or down as the captain commanded. The more Johnson saw of his captain in the control room, the more he respected him. 'He was a man you were bound to respect. You knew he was a man. In fact, your life was in his hands. And he always seemed to get out of trouble. I don't know how we got out of it but we did. He always seemed to know what to do. As the war went on, I got to understand the way he worked. When he was standing there he would turn towards the periscope and I would start to put it up. He sometimes didn't even say "Up periscope". It was like a shrug of the shoulders and it was a message. Whether it was telepathic or not, I don't know, but I used to put the periscope up. I studied him. I studied him intently. He was the one man I would sail to the ends of the earth with. Was there a unique relationship between a crew and captain? I should say. The man brought us back. I'm sure if others had been in charge we wouldn't have got back.'

For Lieutenant J. H. Robinson, *Rorqual*'s Navigation Officer, the tension among the members of the crew at the moment of an attack was tangible. Relaxation only ever came when the first torpedo hit. 'And you waited for it. The captain would have a quick look to see what was going on and if there was an escort pointing towards us he'd decide then what avoiding action to take. The first avoiding action, of course, is 80 feet. Straight down to 80 feet and no problem at all. Or 100 feet. Get well down out of the way. And of course in some places, in the Aegean Sea, particularly off the Dardanelles, there was a lot of fresh water coming into the Mediterranean and you got what were known as layers – a layer of fresh water and a layer of salt water with a layer of fresh water underneath it. And if you could get underneath the freshwater layer it was sometimes a fine defence because sound didn't penetrate for some reason. The only snag of course, was that the weight of the submarine was calculated for salt water and if you hit a freshwater layer, you'd drop like a ruddy stone until you hit the next salt water layer. And you hoped to God there was one within dropping distance.

ABOVE 'Your life was in his hands and he always seemed to get out of trouble. He always seemed to know what to do.' Tom Johnson on Lieutenant-Commander Lennox Napier, shown here looking through the periscope, in battle.

LEFT Off-duty leave for the crew of *Rorqual* in the Mediterranean was very confined. In the early years Tom Johnson (left) and Nick Carter and their mates were limited to runs ashore only in Alexandria and Malta.

'The detection methods the enemy was using were highly developed as far as I know. They had very sophisticated hydrophone equipment, probably as sophisticated as a lot of it today. An escort came after us after we had sunk one tanker and started popping depth charges around us and it was a bit nerve-racking, but everyone I looked at – I'd never been in a depth charge attack before – was looking bemused. One chap had a piece of chalk and a piece of board stuck up against the wall and he was marking down the depth charges. And they'd go away and then they'd come back and there would be another attack and he'd start marking again. I think on that occasion there were forty-four depth charges dropped on us. I may be wrong, but I seem to remember it was forty-four. It didn't quite make the forty-five. Some time after an attack the captain would order the boat back up to periscope depth and he would have a look around and say "Coast is clear. Fall out diving stations. Carry on patrol routine." We were back to normal, back on patrol looking for someone else to attack.'

The relief when destroyer attacks were over was always evident on the faces of the crew, but the number of escape options available to Napier were limited. He could run, but *Rorqual* was extremely slow compared with surface craft; he could dodge, but that required timing and luck; or he could dive, but *Rorqual*'s shape precluded any depth greater than 200 ft (60 m). 'There were restrictions on the depth to which we could go,' he remembered. 'This was because in order to fit the hull to carrying the train of mines inside the upper casing of *Rorqual*, the afterpart of the pressure hull had been flattened. This flattened part tended to distort at even normal operational depths and during the whole of our second period in the Mediterranean we were very considerably restricted to the depths at which we were supposed to go. I think it was only 120 feet. One would have gone below these depths in an emergency, probably with perfect safety, but if it could be avoided we tended to remain at a shallower depth.

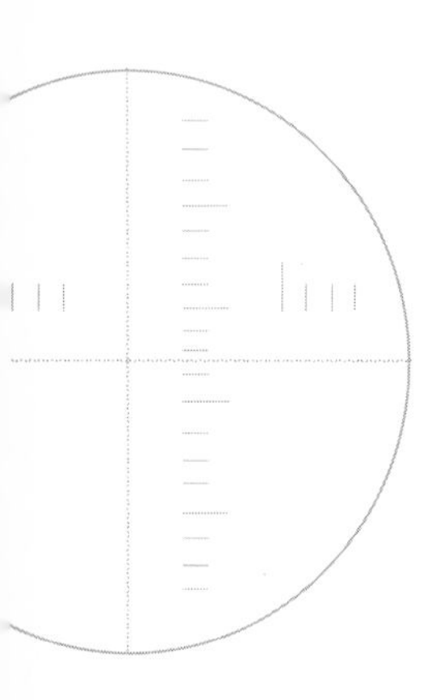

'What successes we had, which were not as great as some others, were, I think, largely due to some degree of caution; not trying to do more than the available ship and force was capable of, and surviving long enough to gather the experience. If you have experience your chances of surviving and carrying out successful operations increases enormously with time. I don't know what the crew thought about these things. I do think they appreciated a captain who had success because it's important to everybody that success should occur. And the longer you go on having success, the more they will come to believe that success will go on. They think, probably, that you are the sort of captain who can be relied upon to have successes without leading them into situations of unnecessary danger.'

Lieutenant-Commander J. P. H. Oakley, who took command in June 1944, was denied the chance to emulate the successes of Dewhurst and Napier. *Rorqual* was ordered from the Mediterranean to the Far East where she saw out the rest of the war. She laid a few minefields, landed an American secret agent – whose real mission was economic rather than military – on the island of Sumatra, and sank three coastal craft with her gun. But for her crew, it was an uneventful tour of duty after her time in the Mediterranean. *Rorqual* left for home in May 1945 and arrived in Portsmouth two months later.

Between January and July 1942 the number of frontline U-boats available for patrol rose from ninety-one to 138. With seventy U-boats now at sea at any one time, Allied merchant shipping losses rose again to an average of 482 000 tons a month, mostly because of sinkings off the east coast of the United States where American convoys of up to ten ships still sailed with full lights as if in peacetime. The U-boats also destroyed convoys in the gulf of Mexico and the Caribbean. In February 1942 the Germans broke the British naval code which allowed them to read most British naval signals for the next sixteen months. They also prevented the British from reading the German codes by increasing the complexity of the Enigma machine. Suddenly, the British were once again blind to U-boat strategy in the Atlantic where, from August 1942, an average of more than 103 submarines was trawling to locate the convoys. However, new radar and anti-submarine direction-finding devices allowed convoy escorts to attack trailing U-boats with pin-point accuracy.

U552: OCTOBER 1942 – FEBRUARY 1944

Klaus Popp was naturally hesitant the day he was due to take over command of the Red Devil boat, as it was known by the enemy. 'To take over a crew which had had such a successful captain as Erich Topp was like taking a dip in ice-cold water. I knew what was going on in the minds of the crew members. What will the new commander be like? How will he behave? How will he cruise our boat? Will he be able to take us back safely?'

Hans Babel remembered the feelings of the crew at the time: 'When Kapitänleutnant Popp arrived on board we noticed that Topp was taller than Popp. I must also admit that Topp was somehow cleverer than Popp. Topp had a great amount of knowledge and was an expert in our field. Popp did not have all these qualities at the start, so it was quite difficult for him at the beginning. Besides, times had changed. We were not hunters any more. From the beginning of 1943 everyone was following us.'

The new commander was confused. His operational orders were to take the submarine and sink ships off the coast of Portugal and afterwards to cruise into the middle of the Atlantic. Popp knew that other front-line U-boats had been ordered to the much more dangerous North Atlantic. Before Topp handed over command, he had explained to the new captain that Admiral Dönitz, the naval officer responsible for U-boat strategy throughout the war, had decided the crew

needed some relaxation in a quieter area. 'After all, they have been on many cruises lately,' Erich Topp told Popp.

The truth was that Topp had asked for, and been granted, a favour by Dönitz: that U552, under its new commander, should not be sent into the North Atlantic again. 'That side of the Atlantic had become a most dangerous area,' Topp remembered. 'I asked, if possible, that Popp be sent to the south side of the Atlantic in order for him to get used to the crew, which was by then a mixture of some old and some new members. Dönitz sent U552 to the South Atlantic and it survived another four cruises.'

Popp got to know his men as they cruised through the Bay of Biscay, one of the most dangerous of submarine routes because of its exposure to patrolling aircraft. 'The first cruise turned out quite well,' he said, 'because nine days after leaving port we sank an English corvette. This gave the crew more motivation and I was assured that they now had respect for their new commander even though we had been chased by two other English corvettes and only just made it in getting away. Afterwards we cruised into the middle Atlantic only to find that big ships or groups of ships had already been in the area and cruised on to Cape Town.

'Our cruise turned out to take ninety-six days and was quite exhausting. The temperature in the engine-room was 64°C and in the boat in general 36°C. It was really hot. At that time the Allies had just landed in Morocco and I was asked if we could cruise to that area. But some of the crew members were suffering from itch or scabies and we also had other illnesses on board. I asked for permission to return to base because of our circumstances on the boat.

'We had sunk one corvette and a cruiser, whose captain we captured. We had orders at the time to intern captains and engineers.'

After ten weeks in port the U552 had been completely serviced and was ordered into the North Atlantic. 'It was very stormy and couldn't be compared to our previous cruise,' Popp remembers. 'We received news that a convoy was on its way home and that we would probably miss it as they had just left America. We cruised ahead under full power but we were hit by a hurricane and didn't even see it.'

One of the look-outs Popp relied on was Emil Lehman, whose duty position on the bridge of U552 was at the front on the port side. He was also responsible for the concentration and efficiency of the other look-outs. 'The difficulty of the job depended very much on the weather. If the weather is nice and bright, there

Germany produced more and more U-boats as the war progressed. This is an assembly line for the Type VII submarine in 1944.

is no problem really. Sometimes you don't even need the binoculars – you can see enough without them. But when it's raining, it could be a strain on the eyes. It's exhausting because you are constantly cleaning mist and water off the binoculars. After being on duty for two hours you start seeing shadows on the horizon and the sea-water starts to burn your eyes. I would be hoping to see a convoy or a single ship cruising on its own because those were our targets. But the longer the war went on the more difficult it became. During the latter part of the war we had to protect ourselves first. We had to watch out for being attacked ourselves. Attacking enemy ships became a second priority. We became very cautious because we knew of the big losses. We knew how many boats did not make it back. It went as far as crew members reminding each other how dangerous it was and saying to one another, "Just watch out up there". It was not a case of getting ready for attacks any more. It was more a case of protection and surviving.'

Towards the end Bernhard Gutschow recognized the increasing fear among the crew of U552. 'Of course, our fears grew but at the same time we were a crew which could be relied upon. Somehow it was worse than being at the front. New people arrived constantly who knew nothing and had to be told. Well, it was all new for them. They had no experience except the one training cruise into the Baltic. Everything had become different by then. New things appeared all the time. Many boats were lost. You'd often hear people commenting that a certain boat hadn't made it back. Of course, being threatened by air-raid warnings, one had to protect oneself, what one had left. When we were under depth-charge attack we knew the men in the machinery room would spot leaks at the right moment. The torpedo mechanic, the radio operator, all could rely on each other. There was always the feeling – "We'll make it again this time". Of course, there is always the feeling of fear as you sit down there, depth charges falling around you and all you get are orders. What's going on? You can only hear screaming, you can't see any-thing. Yes, and the fear grew and grew as the attacks were on the increase. But this fear did not lead us to do anything we didn't want to do. We didn't lose control. It was very bonding. We didn't have time to ask ourselves, "What is all this for?"'

Kurth Kraus was responsible for operating the machinery that balanced the trim of U552. On command he caused water-ballast to be pumped fore or aft to keep the boat level. If the alarm went when the boat was on the surface it was his job to pull the valves that flooded the reservoirs and hold on to them until they were fully flooded. 'I was very frightened when destroyers launched depth charges,'

A torpedo hit amidships. Another enemy vessel sunk. Topp's crew watch the destruction of a tramp steamer.

he admitted. 'The worst, though, was the attack of an aeroplane at the very end. One bomb fell at the stern so the bow went up. Then it happened the other way round, a bomb hit the bow and the stern went up. We couldn't see anything and couldn't help. We could only rely on our commander to do it right.

'We dived at about 45 degrees in a sloping position. It was total confusion. Hardly anybody could manage to keep themselves on their legs. When we got to nearly 150 metres we tried to balance out. We had to get the bow up and the stern down to make sure that the last air-bubble would get out of the dive cells. Then they could be closed again. Another depth charge fell pushing us down further. I heard a voice: "Boat at 160 metres, boat still falling, 170 metres ... 180 metres". Then I saw my colleague behind me putting on his life-jacket. It had a little bottle of oxygen which could keep you under water for about half an hour as we had been taught at diving school. What is he doing? I remember thinking; at this depth all is lost. Then I heard, "200 metres – boat still falling". Is this possible? I thought to myself, we have no water breaking in. It was frantic in the engine-room, they had their hands full dealing with the damage from the depth charges. Everything was working, the motors and both the hydroplanes, but the boat still sank.

'I must have had a guardian angel. Suddenly, at 220 metres, I walked past the tower. I remember seeing 223 metres on the depth gauge. There was a valve near the gauge. When water comes out of this valve it means the reservoirs are full. To allow us to surface they should have been empty. I turned the valve and water came out. I shouted to the engineer. He looked shocked and shouted, "Blow out". Nobody had thought about it – everybody was trying to hold on to something and thinking "This is the end". Then, to everyone's relief, we heard the person responsible for the hydroplanes say, "Boat steady at 230 metres". At that moment I felt like it was my birthday.'

In early February 1944, U552, damaged during the attack, was sighted by a destroyer. 'We realized too late that he had seen us,' Popp recalled. 'We got attacked just as we wanted to dive. Nursing a damaged submarine, we were unable to dive away quickly. We could only dive down to 30 metres. Meantime the destroyer had set its bombs to hit. Luckily we were only floating near the surface and didn't get hit by the second attack because the destroyer had set its depth charges to explode in much deeper waters. So, we made it back again, with major defects and some crew members injured. But we made it.

'We should have been arriving back with five other boats, but out of those five only two made it back, the other three never returned. In situations like that you would ask yourself if it was to be your turn next. It was a matter of sheer luck that our boat had made it through the war. Luck was always on our side.

It could have been different. We escaped a bitter end. It always was my inner wish, my deepest wish, to take the boat back safe and sound with all its people. But that, of course, is only a dream. What I mean is that one cannot behave in that way. Risks had to be taken. My point of view is that maybe we only took those risks when they were properly calculated and that's why we lived through the war uninjured.'

The submarine blockade of the Japanese home islands in the summer and autumn of 1942 had been effective. On one of her patrols off the Japanese coast near Osaka, Silversides *sank transports totalling 10 000 tons on 28 July and 8 August. By the end of the year the Japanese had been defeated at Midway, stopped in the Aleutians and pushed back in the Solomon Islands. As the year turned there were more than fifty fleet submarines in the Pacific and Japan had lost scores of merchant ships and several fighting ships. On 18 January 1943, Burlingame attacked and sank the* Toei Maru*, a 10 000 ton tanker, and, the following night, three freighters with one salvo of five torpedoes. A sixth remained stuck in its tube where it might have exploded until Burlingame put the submarine into reverse and ordered the torpedo to be refired.*

Silversides *set out for base two days ahead of schedule. While she went for a refit, the number of submarines blockading the Japanese home islands between April and June was three times the number patrolling there in the first three months of the year. The January patrol had been Burlingame's last in* Silversides.

Silversides: JULY 1943 – JUNE 1944

Lieutenant Robert Worthington thought the new captain of *Silversides* was almost the antithesis of Creed Burlingame. Lieutenant-Commander Jack Coye was relaxed by nature. However, like Burlingame, he was gung-ho to get at the enemy. 'We all loved Creed Burlingame but his successor turned out to be an excellent skipper. He was very calm and deliberate – and he was very considerate and prone to put faith in his officers and men. I had been with *Silversides* since the beginning and so I felt like she was my ship. I knew more about her than any man aboard, but Jack Coye and I had no problems in accommodating each other.'

Coye himself remembered feeling extremely fortunate to be given command of *Silversides* when he first boarded at her Pacific base in Brisbane, Australia, in July 1943. 'Burlingame had done so well and was so outstanding that he was a real tough act to follow. I felt I had to try to keep the crew organized as they were and not change anything, and hope that I could fit in and try to do it somewhere near as well as he could.

'Some men who were captains of submarines when the war started were eventually relieved of command because they weren't aggressive enough. They got scared too easily, I guess. But we, the younger ones, were more aggressive.

RIGHT Lieutenant-Commander Jack Coye took over command of *Silversides* in July 1943. 'Creed Burlingame had done so well. He was a real tough act to follow.'
BELOW A familiar view from the conning tower of *Silversides* as she hunted for Japanese targets in the Pacific in 1943.

We didn't know any better, I guess. But the normal procedure was if you didn't perform in one patrol you would probably get a chance at another one. But if you weren't aggressive in the second one they would relieve you and put another guy in. So you had to do well.'

And under Coye, *Silversides* was again extremely successful. 'On the next patrol we had a very successful run and we sank at least four ships, maybe, and damaged some others. I preferred to make the majority of my attacks at night. We would normally sight a convoy in the daytime, usually from its smoke, and then we'd get up ahead of them and wait until it was night. If we had to attack in the early morning we'd do it by periscope. That's in general how we did it. We were very fortunate because in a couple of attacks I got three ships at a time. I was lucky.'

Gene Malone, the radar officer on *Silversides*, remembered the lack of commands during those attacks. 'A submarine, by its nature, is almost a perfect example of teamwork. You didn't have to think about what anyone else was going to do or what was going to happen because it was going to happen as a piece of absolute perfect teamwork. Captain Coye and I rapidly formed a close team relationship. There were no direct verbal communications. It was a matter of being in each other's minds and working together almost as one. I think it was best expressed by Captain Coye on one occasion when we were in a very difficult situation on the surface. I was feeling we should be firing fairly soon and he was feeling we should attain a better firing position. Afterwards we agreed that we came to our conclusions when he on the bridge could hear my knees knocking in the conning tower and I, in the conning tower, could hear him burping on the bridge.

'Coye's mind was never still, it was always alert. I don't think people realize that fighting with a submarine is essentially an intellectual process which takes a lot of guts, a lot of bravery. You have to ride into danger, but if you're going to be successful, you never do so without knowing precisely what you're doing and precisely what the odds are. Unless you have that intellectual approach you don't come home. Jack Coye communicated that sense of intellectual competence; of intellectual command. At the same time he was communicating a sense of essential fearlessness: "Hey, fellas, everything's OK. We're going to kill these guys". Wonderful man to work with.'

Coye, also, remembered the times when *Silversides* was hunted by the Japanese. 'Of course the object was to get away from the destroyer, so you tried to put him aft and open out from him and you went deep. Our test depth was 300 feet but sometimes we went a little bit deeper. And you had to run slowly, otherwise your propellers would cavitate, and that would help him. But then, when he did drop his depth charges, you could speed up because that would hurt his sonar gear and you could, maybe, get out of the way a little bit. And you could tell the distance of the depth charges. Before you heard them there'd be a click. The click was the pressure wave coming and that was what really hurt the submarine. The bang didn't really hurt, it was the click. There was a time interval between the two and the number of seconds you could count indicated how far away the depth charge was. When the clicks and the bangs got together, it was getting serious and you had to do something. In the early part of the war we didn't have much problem because the Japanese weren't setting depth charges very deep. At a depth of 300 feet we were pretty safe. When the Japanese began setting their charges a lot deeper the United States began making submarines that would test up to 400 feet.'

Worthington remembers that by the end of Coye's tour of duty in early December 1944 they were struggling to find targets. 'From our refit base at Pearl Harbor to the war zone was two weeks either way and there were no enormous hundred-ship convoys – the kind the Germans used to get in the midst of and then sink half a dozen ships in short order. Convoys, in the kind of areas we were attacking, consisted only of two to four ships, maybe six at the most.'

Gene Malone and the other members of the crew felt equally frustrated. 'The last patrol I made in *Silversides* was the eleventh. It was in late 1944 and at that point the merchant-type targets were becoming quite scarce and were being very carefully protected. They were running close inshore along the China coast and weren't going out into the open sea any longer. So those ships were hard to find. The major warships were also being protected so they were also hard to find. It was definitely a thinning out. By that time we had sunk so many of their merchant ships in the open ocean that there weren't too many left.'

When Coye passed command of *Silversides* to Lieutenant-Commander John C. Nichols he had no regrets. 'I felt confident in turning *Silversides* over to him because I'd known him for a long time and he deserved a chance. I felt a little sorry because I knew that there weren't that many targets left out there. Most of them had been sunk. But *Silversides* rescued some aviators and did a good job and he brought it back. Yeah. That was most important.'

The Americans in the Pacific, the Germans in the Atlantic, the British in the Mediterranean all used the submarine as a strategic weapon to destroy the ability of their enemies to supply their armies and their civilian populations. The Americans wiped the Japanese merchant fleet off the surface of the Pacific. The British in a tiny, hostile war zone had only limited success; the German submarine threat was eventually overcome only by a mixture of scientific innovation, code-breaking, sheer weight of opposing forces and the rest of the war going badly on the mainland of Europe. The lesson was simple: the submarine on its own can never win a war. But only Frank Jordan, Bernhard Gutschow, Robert Chandler and the other men who, like them, served in Rorqual, U552, Silversides *and every other submarine in the Second World War, will ever really be able to speak to each other in a common language to discuss how close to victory they ever came.*

CHAPTER 7

NOWHERE TO HIDE

The Italian divers rested. They switched off their underwater welding equipment and examined their work. The hull of the scuttled oil tanker *Olterra* stretched away from them into the clear blue depths of the harbour in the neutral Spanish port of Algeciras. The divers had been working secretly there for months. From the surface they knew the *Olterra* would create no more interest than any other half-sunken hulk in any one of a dozen harbours around the Mediterranean. The fact was, though, that it was not anywhere in the Mediterranean but directly across the bay from the British naval stronghold of Gibraltar where look-outs constantly monitored activity in the port. The Italians had ensured that the flames and sparks of their welding torches would not be seen by rigging up a screen of blankets around them as they worked in the water. And now their task was finished. The two hinged flaps they had been cutting into the hull of the scuttled tanker were invisible to anyone who did not know where to look. In the autumn of 1941, the only people who would have known this were the handful of divers who together with their human torpedoes made up the *Decima Flottiglia MAS*, the 10th Light Flotilla of the Italian navy, which had been attacking Allied ships inside safe anchorages in the Mediterranean for the previous fifteen months.

Italian naval officers and engineers of *Decima MAS* had arrived at Algeciras that autumn with the brilliant idea of turning the scuttled *Olterra* into a secret base from which they could attack shipping at anchor either inside or outside Gibraltar harbour. The Spanish authorities had been told that the team of salvage experts was there to repair the vessel so that it could be sailed again once the war was over. By December, however, its hull had one opening to allow men to enter and another, 25 ft (7.5 m) wide, to launch and receive the two-man torpedoes the Italians were using in their attacks on Allied shipping.

'The Spanish were totally unaware of what was really going on,' Ernesto Notari, one of the Italian officers recalled. To get the torpedoes and the rest of the support equipment from their home port of La Spezia to Algeciras, officers of *Decima MAS* needed official export clearance – but the operation was entirely secret. A few of them broke into the Italian Foreign Office in La Spezia one night and some days later all the cases destined for Algeciras had been marked with the official seal they had stolen. 'From La Spezia, two train wagons left with regular permits,' remembered Notari. 'They were carrying gear officially for the refurbishing of the tanker, copper pipes and things like that, but inside the big containers were hidden all the parts of the submarines, diving suits, breathing apparatus, oxygen cylinders and explosives.'

The first attack on shipping in Gibraltar harbour from the *Olterra*, on 7 December 1942, was disastrous and it took several months to replace the equipment lost in that raid. The second, the following May, was a brilliant success. Three human torpedoes, led by Notari, left the *Olterra* and headed for the Bay of Gibraltar. There they sank three Allied merchantmen, totalling nearly 20 000 tons, before heading back for the safety of the *Olterra* where all the explosions had been clearly heard.

Scuttled, rusting, abandoned in Algeciras Harbour, the *Olterra* was turned into an underwater base from which Italian frogmen attacked ships in harbour in Gibraltar across the bay.

Three months later, the human torpedoes slipped away from the *Olterra* once again and a further three ships were sunk. This time they totalled 23 000 tons. But only five men made it back to base. As Notari was fixing the explosive charge to the 7000 ton American-built liberty ship *Harrison Grey Otis*, his Number Two, Andrea Giannoli, developed oxygen poisoning and swam for the surface. He left Notari hanging on to the torpedo, which was out of control. It dived more than 100 ft (30 m) then turned to head straight for the surface again at great speed, emerging only 3 ft (1 m) from where the warhead had been placed on the side of the ship. Amazingly, no one heard the tremendous noise the torpedo caused and Notari moved away from the area without being seen.

Few people would have believed that *Decima MAS*, born of a group of idealistic friends, would be capable of such exploits. It had all begun because of the fear submarines had spread among the admirals responsible for the safety of the capital ships of their respective navies. The destructive power of the craft had forced the battleships of many nations to shelter behind anti-submarine nets and mined channels for most of the First World War. The Austrian fleet, based at Pola on the Adriatic coast, had been no different.

At the end of the war, two Italian naval officers, Raffaele Rossetti and Raffaele Paolucci, decided to strike at the Austrian enemy inside Pola harbour. They had been working independently and Paolucci had designed an explosive charge while Rossetti had designed a carrier for such a charge using the shell of a German torpedo. When they got together they created a compressed-air driven vehicle more than 29 ft (8 m) long and weighing 1½ tons. Its detachable magnetic warhead, containing 375 lb (170 kg) of explosive with a clockwork mechanism to control the time of explosion, could be clamped to the side of a ship. The two-man vehicle was given the name *Mignatta*. On 18 October 1918, after negotiating every obstacle to get to the 20 000 ton Austrian battleship *Viribus Unitis*, Rossetti and Paolucci clamped their warhead to its hull only to find, just before the explosion damaged the ship, that the war had ended and that the ship had, for a few days, been the pride of the newly formed Yugoslavian navy.

When Mussolini invaded Abyssinia in 1935, two naval engineers, Teseo Tesei and Elios Toschi, both sub-lieutenants, proposed a manned torpedo based on the *Mignatta*. By January 1936 the two officers had successfully taken it through its tests, and as the world moved towards full-scale war, Tesei and Toschi began to improve the performance of their craft. Early in 1940, in the Gulf of Spezia, three *Maiali* were launched from the submarine *Ametista* by its commander Prince Junio Valerio Borghese, and one of the divers succeeded in attaching a dummy charge to a ship in harbour. As a result, twelve more *Maiali*, 5 ft (1.5 m) shorter than Rossetti's *Mignatta*, were ordered. The compressed-air power unit had been

The torpedo vehicle codenamed *Maiale* – pig – which the Italians used to attack Allied shipping at Malta, Gibraltar and Alexandria.

changed for an electric motor which gave near-silent running over a maximum range of 5 miles (8 km). The crew wore rubber suits and had a six-hour supply of oxygen for their self-contained breathing apparatus which allowed them to take the craft down to 30 ft (10 m) over a short distance. The explosive charge was 485 lb (220 kg) and could be set for any time up to five hours ahead.

Unlike most designers, Tesei remained involved with the operation of his craft until the end. One of his colleagues, Gino Birindelli, remembered that the formation of the 10th flotilla was based on a number of schoolfriends including himself and Tesei who had attended the Naval Academy together in the early 1930s. 'The craft which had been built for the Ethiopian war were put in a workshop in La Spezia naval yard and forgotten. Then something else happened. Tesei, Toschi, myself and others, all old friends, were officers on submarines and at night, when we were not at sea on exercise, we used to go to a restaurant and we started talking. And of course we were young men and we were strong nationalists and we were thinking of Europe. What's going on. Hitler. Germany. There's going to be a war. And so, in 1938, when nobody else was thinking about it, we started thinking again about Tesei and Toschi's craft.

'In September 1939, when war broke out, we wanted to find a place which was far away from everything and the Duke of Salviati allowed us to use part of his estate. It was completely secluded. There were pines, there were wild horses, there were deer. Life was a dream. We were alone. We had nothing. We bought stoves with our own money. We used to cut the wood to warm ourselves. We improvised a shower. We just lived out our dreams of winning the war and practically nothing else. We didn't have the machines but we started training by using breathing apparatus and then we began to develop the proper suit. It was supposed to be easy to wear but it leaked and it really tormented us.

'Then the torpedoes were brought out of storage from La Spezia and we began to practise with them. How to dive, how to come out of a dive, how to steer and how to get through a torpedo net. So it was really very complicated. And then we had to approach a ship and find a way of attaching the explosive warhead of the torpedo to the hull. But everything we needed made had to be made in La Spezia and we weren't allowed to approach them directly, we had to go through Rome. It was all very secret. We had to have a compass that could be seen under water at a depth of 100 feet at night and we had to have maps and watches which could be read under water. And we had to develop the explosive head, the time trigger, the fuse, and the batteries. And we didn't have the technicians who could make a drawing so that the workers at La Spezia could make what we had planned,

The group of friends at their secret training place. Gino de la Penne is seated far left, with the dog; Borghese is standing, fourth from the left.

and on the estate we had no ship on which we could practise. It was really extremely difficult. The only ones who thought we had a workable idea were ourselves. The others didn't know anything. It had to be a secret.' It was from these days on the farm that the torpedo was given the name *Maiale* – Pig.

By August 1940 they were ready. The friends had intended their first raid to be on British capital ships anchored in the Egyptian base of Alexandria but, as the torpedoes were being transferred at the start of their journey, three British Swordfish bi-planes attacked and sank the submarine that was to act as tender for the two-man craft. The very first operation was a total failure. Apart from the submarine, forty-five men were lost.

The failures continued. A *Maiali* operation against Gibraltar in the late summer of 1940 was aborted and when Birindelli got back to base he was told that a parallel operation had ended in disaster when the tender submarine had been destroyed. Some of his closest friends, including Toschi, had been killed. 'We went back to our secret base', he said, 'and at a meeting there we came to the conclusion that having twice tried to get near to the enemy and having lost two tender submarines, probably our ideas were not exactly right. So the decision was that we would try again, this time against Gibraltar. If the attack was successful, the *Maiali* would be employed again. If the attack was unsuccessful, the *Maiali* had to be forgotten and put aside because it was too difficult, too dangerous and too costly.'

They decided to attack on the night of the new moon in October 1940. The submarine *Scire*, commanded again by Prince Junio Valerio Borghese, carried the *Maiali* to within range of Gibraltar harbour. Three were launched, led by Tesei, Gino de la Penne and Birindelli, who remembered saying to his friends: 'The British say that no gentleman goes out without a shave. Let's shave. We must see the British perfectly shaved. So we shaved'.

The battleship HMS *Renown* was allocated to Tesei. An unknown cruiser was allocated to de la Penne and Birindelli was to sink the battleship HMS *Borough*. The cylinders containing the *Maiali* were released from the *Scire* and sank to the sea-bed. The three commanders had agreed to meet on the surface twenty minutes later to begin the journey into the harbour. Birindelli and his partner, Petty Officer Damos Paccagnini, took longer than that to release their *Maiali* and when they reached the surface their colleagues were nowhere to be seen. Birindelli assumed they had gone on ahead. But he was not worried. 'The general idea was that we would attack the British shipping and then go to the beach in Spain where some intelligence agents would meet us to take us to Italy so that next day, at one o'clock, we would have lunch in the Navy Club in La Spezia. We wanted the British to have seen their ships sunk and not be able to understand

Gino Birindelli, one of the founders of the group which perfected the technique of underwater attacks on Allied ships.

who had done it because even if by then they had some idea of the *Maiali*'s existence, we would be in La Spezia having lunch.

'So I started going towards Gibraltar and I found myself in between two rows of merchant ships and tankers. We went right through this double column. I could see the men on deck, smoking, talking. It was unbelievable. Then I got to the harbour. I couldn't go through the main entrance because I knew it was tightly closed. I knew there was another entrance, but it was very shallow and I knew that there were nets, so I had to go through the only entrance remaining. As we had planned it I would go under the nets, emerge, get into the harbour and look for my target.'

But water was leaking into Birindelli's craft and he could not submerge so he had to improvise. 'I said, all right, I'll go over the booms. But the entrance of Gibraltar is 50 metres long and there were sentries on the piers and I went over the booms with this big thing, 7 metres long, one ton in weight and nobody saw me. By that time I was terribly tired. I couldn't even think much. I just moved automatically. It was three or four o'clock in the morning and so I went through the first booms, over the second and then to the left was HMS *Borough* 300 metres away. A big chunk of steel. Dark. Immense. Unbelievable. What can I do? What about de la Penne and Tesei? What happened to them? Then I said, all right, they are far ahead of me. I don't care because they have already carried out their attack. They are already in Seville. I'll carry on. By that time I was inside the harbour which is 14 metres deep. So I just sank my hull and went to the bottom and started moving toward the target by paddling along on the bottom.'

For technical reasons Birindelli was unable to attach the explosive warhead to the hull and he decided to place the entire *Maiali* under the keel of the battleship. But he was still 76 yds (70 m) from the ship. 'Can you imagine, after having gone through all these things you have to stop? By this time Paccagnini could stay down no longer and had gone up to the surface. And I started wrenching, pushing, pulling. I felt I must get there. The breathing apparatus didn't work any more and I felt that I was passing out. At this moment I just had time to start the timer for the explosion and then I got to the surface and I said to myself, "They did not see me when I got over the booms and I am here, 50, 60 metres from a battleship with 1500 men on board and they didn't see me." So slowly, slowly, moving slowly, I went over them, I took off my suit and tied it under one of the buoys.

'I threw off my breathing apparatus and I started swimming towards the Spanish coast but I got cramps, terrible cramps. So now I am sinking because I cannot move. And I moved towards the pier where there was a steel rope hanging down and I got hold of it. I next remember lying on the ground on the cold pier with nobody there. I took off my watch and threw it in the water because no one must know what was happening. I was on Gibraltar pier at four o'clock in the morning and I couldn't believe I had got there. And I started moving. Then I got to a place where the pier was getting narrow and there were some British there and I remembered I couldn't pass in front of them because I was completely drenched. Then a boat came alongside and moored and I was there but they didn't see me. Then I went on and got to the town pier and it was already daylight and I said what am I doing. Just jump on the pier and move. Try to get somewhere. I got on the pier and started walking and I think the respect for privacy of the British helped me because nobody asked me anything. All around me there were sailors and workmen.

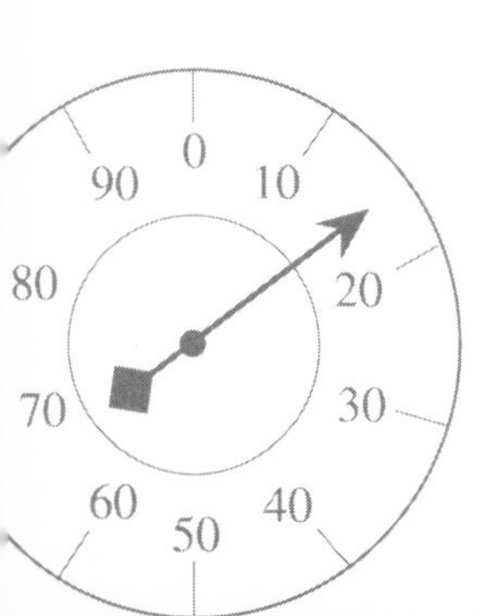

'I got to the end of the pier where you either go into town or to the other merchant pier and I saw there was a small ship there with a Spanish name. I went there and got inside and told a Spanish sailor I wanted to stay there until night-time. But two sailors had seen me getting on board and took me for questioning. And while I was being questioned there was a huge explosion from the harbour.'

At that moment Birindelli believed the operation had been successful, and that the *Decima MAS*'s ideas were good. In fact, the explosion had damaged none of the targeted ships. Nevertheless, a secret message from Birindelli encouraged the *Decima MAS* to continue its operations.

More than a year later Gino de la Penne and Petty Officer Diver Emilio Bianchi led a team of three *Maiali* into the harbour in Alexandria where the British battleships HMS *Valiant* and HMS *Queen Elizabeth* were riding at anchor surrounded by their own anti-submarine nets. A few hours after the two-man teams had fixed their charges, both battleships were seriously damaged and settled on the bottom of the harbour. For the crews of the *Maiali*, who were captured immediately after the raid, the war was over; for the Italian nation, this action against the majesty of the British navy restored self-respect; for Winston Churchill, the Italian underwater commando campaign was intolerable.

'Please report what is being done to emulate the exploits of the Italians in Alexandria harbour and similar methods of this kind. Is there any reason why we should be incapable of the same kind of scientific aggressive action the Italians have shown? One would have thought we should have been in the lead. Please state exact position,' he wrote one month later.

The fact was that until the Birindelli raid at Gibraltar the British had little information about the technology of manned torpedoes. As a result of that raid, two naval officers got the chance to examine, from a range of 60 yds (55 m), the *Maiali* crewed by either Tesei or de la Penne. Neither man reached Gibraltar harbour and one of their craft was washed up on the coast of Spain.

Skilled interrogators also persuaded Paccagnini, who had been found in the water, to talk to them about the *Maiali* and to confirm details of a sketchy outline they had put together from the brief sighting by the two officers in Spain. Birindelli said nothing. Both men spent the rest of their war as prisoners of the Allies.

The British worked furiously to create their own manned torpedo to strike back at the Italians. However, by the time they had perfected such craft, referred to as *Chariots*, the war in the Mediterranean against the Italian navy was almost over. Apart from one successful raid on a light cruiser and a merchantman in Palermo harbour at the beginning of January 1943, and two others at La Spezia the following June when one heavy cruiser was sunk and another damaged, the British initiative had no real impact in the Mediterranean.

An X-craft midget submarine under way.

By contrast, the X-craft, the British midget submarine that was being developed at the same time as the *Chariot*, achieved perhaps the greatest success of the entire campaign.

For once, Winston Churchill, impatient and tireless in his demands for action, had been anticipated when he again minuted his service chiefs in February, 1943: 'Have you given up all plans of doing anything to *Tirpitz*…? We heard a lot of talk about it five months ago … It seems very discreditable that the Italians should show themselves so much better in attacking ships in harbour than we do … It is a terrible thing that this prize should be waiting, and no one able to think of a way of winning it.'

The German battleship *Tirpitz* displaced 52 600 tons and with eight 15 inch (38 cm) guns was an ever-present threat to the Allied convoy routes to Russia from her base in Altafjord in northern Norway. On six occasions Bomber Command had tried to sink her and had lost twelve aircraft without even damaging the ship. And in September 1942 weather conditions had wrecked an attempt to take two *Chariots* to the fjord to attack the battleship. But in June 1943 training began at Loch Cairnbawn in northern Scotland when the commanders of six midget submarines were initiated into the group organizing 'Operation Source': the plan to sink the *Tirpitz*, the battle cruiser *Scharnhorst* and the heavy cruiser *Lutzow*.

X-craft X-5 to X-10 were allocated to the mission. Each displaced nearly 30 tons and was 51 ft (15.5 m) long and nearly 6 ft (2 m) in the beam. The vessels had a surface speed of 6.5 knots generated by a 30 horsepower electric motor, a maximum range of 1500 miles (2400 km) when fully loaded and carried a crew of four, one of whom was a trained diver.

The craft had no torpedoes but each carried two large, delayed action mines fastened to its sides and had a 'wet and dry' compartment which allowed a diver to exit and enter the submarine in order to place these limpet mines against the sides of enemy ships or deal with underwater obstacles. Unlike the expendable *Chariot*, the X-craft was also designed to get back home after a raid.

On the night of 11–12 September 1943, the six X-craft, each towed by a standard submarine, set off for the eight-day journey. They were at least 20 miles (32 km) apart and were heading for Soroy Island 11 miles (18 km) from the Norwegian coast and 100 miles (160 km) from their target at anchor in Kaafjord, at the very top of Altafjord.

X-5, X-6 and X-7 were to strike at the *Tirpitz*; X-8 at the *Lutzow* and X-9 and X-10 at the *Scharnhorst*. X-9 was lost *en route* and X-8 was damaged by a premature explosion and scuttled. At first light on the morning of 20 September, the four remaining craft left Soroy Island for the next rendezvous point off Tommelholm Island near the entrance of Kaafjord. X-10 had leaks which put her periscope and gyrocompass out of action and she took no part in the final attack. As it happened both the *Lutzow* and the *Scharnhorst* were away from their anchorages and the remaining X-craft were able to concentrate their attack on the *Tirpitz*.

The diver on X-6, commanded by Lieutenant Donald Cameron RNR, was Sub-Lieutenant Richard Kendall. He was aware of the theory but knew it was going to be different in practice. 'The theory was that when you came to your first net you had to go in at about a depth of 30 feet. The captain would put the nose of the submarine right into one of the diamond-shaped meshes of the net. You couldn't go any deeper than that because anything deeper and the diver was likely to get oxygen poisoning. Most of the nets were 3 foot diamond mesh and the diver went out and stood on the casing and cut through two of the sides

The crew of *X-6*. Lieutenant D. Cameron VC, RNR (centre, back row); Lieutenant J. T. Lorimer, RNVR (right, back row); Sub-Lieutenant R. H. Kendall, RNVR (left, front row); Eddie Goddard (second from left, front row).

of the diamond on the top. Then the submarine could be pushed through part way and then you'd cut another one where it was likely to be caught and then the diver would ease the submarine through until the net went over the end of the rudder. And once you saw the net disappear behind you, you then rushed back as fast as you could and got in through the wet and dry compartment again and slammed the hatch down.

'However, on the actual night, the captain looked through the periscope and saw a boat approaching the net and the guard vessel was opening the net for the craft to go through. He came up on the surface and went through the opening immediately behind the boat. He then crash-dived and got out of the way.

'But we had our problems. Our periscope had flooded and more often than not we weren't be able to see through it. It was a bit like Jack-in-the-box. We would come to the surface and Lieutenant Cameron would not be able to see anything. So he would go down and stop on the bottom for fifteen minutes at a depth of 80 or 90 feet and dry out the periscope from inside and come up again and have a quick look. Unfortunately every time we did this something occurred. On the first occasion we were about 30 feet from a boatload of sailors so we had to go down again. The second time we were passing between a moored tanker and her anchor chain, so we had to go down again and there really wasn't time to do anything else before we bashed into the nets around *Tirpitz*.

'By this time we were very late and we went down to the bottom because torpedo nets are only supposed to go down 30 feet. But the Germans had cheated and these all went down to the bottom. So we were rather upset about this and tried to get underneath at various places, more or less scraping along the bottom. By now it was daylight, but we didn't try and get through by cutting because the mesh of torpedo nets hanging around ships is very, very small indeed. It's probably only about 18 inches square. Eventually the captain gave up the idea of getting under the torpedo net and came out astern from it. He came up on the surface to have a very quick look. There he found a low spot in the nets and went hard at it on the surface. This was at about seven o'clock in the morning. It was fully light so we went over the top of the net and immediately submerged. We had basically arrived. All we had to do was carry out our attack with both mines fused to go off an hour after they were dropped.

'The idea was that we would go along parallel to the *Tirpitz* and then turn through 90 degrees and drop one charge under her bridge, then turn another 90 degrees and drop the second one under her screws. Well, we were just about to make our first 90 degree turn when we hit a rock. We came full astern off that rock and turned through what we thought was 90 degrees. But hitting the rock had wrecked our compasses. So we went towards where we thought the *Tirpitz* was and, er, we missed her and hit the net on the other side. We then came astern and missed her again. And then we didn't know where we were at all. So we eventually came up on the surface.

'We looked through the scuttles and could see the *Tirpitz* sitting there, about 150 yards away. The Germans by this time were well awake and were firing machine guns and hand grenades at us, which weren't actually doing us any harm. Then we went full ahead and as we got very close to her, within about 20 feet, we released both mines so that these would float down under the target. The only thing to do then was to scuttle the craft on top of our mines so that when they went up they would destroy the craft. There didn't seem to be any point in going down with the craft, that wasn't going to do anybody any good, so Lieutenant Cameron set all the controls to pump water into the boat and sent John Lorimer, myself and Eddie Goddard out with our hands up. As the craft began to sink he came up too and they took us all off on a launch.'

Within half an hour Kendall was on board the *Tirpitz* looking at his watch. 'Well, of course, we had about 4 tons of Amitol underneath so there was likely to be a big bang. We had no idea of what was going to happen. We just stood there. I was concerned that at any moment there was going to be a ginormous explosion. The fuses would be activated by any other explosion close by and we knew there were two other craft aiming at the *Tirpitz* so there might well have been something like 12 tons of explosive under her which would go up at the same time. Strangely enough, the actual explosions were not enormous. There wasn't very much noise. I was thrown off my feet and the *Tirpitz* took an immediate list to starboard. There was an enormous feeling of exhilaration. We'd done what we meant to do. I felt tired at that stage. I don't think I felt frightened. There was this overriding feeling of success. Obviously we hadn't sunk her but we'd done a great deal of damage.'

As X-6 came under fire from the *Tirpitz*, X-7, commanded by Lieutenant Godfrey Place RN, was scraping along the side of the battleship at a depth of 40 ft (12 m) placing one charge forward below the keel, and another near her stern. As the X-7 attempted to get away she became entangled in netting and was still there when the mines exploded under the *Tirpitz*. The shock waves from the explosion damaged her so badly that Lieutenant Place was forced to surface and surrender. As he did so, his craft sank, trapping and killing two of his crew.

Tirpitz was out of action for six months until March 1944 and was never again fully operational. She was eventually sunk by RAF Lancasters of 617 Squadron on 12 November 1944. The X-5, the only X-craft unaccounted for, was both reported to have been sunk in the fjord by gunfire immediately after the explosion and also to have been glimpsed leaving the fjord on the day after the attack. She was never seen again. The crippled X-10 got into difficulties on the homeward tow and was scuttled. All six X-craft involved in 'Operation Source' had been lost.

Although *Maiali* and *Chariot* crews were always unlikely to escape from the scene of their activities to fight another day, there had never been any question of the manned torpedo becoming literally a suicide craft. In 1944, however, that was the situation that developed in Japan.

The Japanese had used midget submarines at Pearl Harbor, where they had been totally ineffective, and, again without great success, during the battle of

'It was a great, big, black thing like an iron coffin. I wasn't scared.' Pilot Officer Ternyoshi Ishibashi on seeing a *Kaiten* suicide torpedo for the first time.

Midway. They had also struck against warships at anchor in Madagascar and in Sydney harbour during the summer of 1942. Their success was limited but in 1944 they turned to the ultimate definition of human torpedo. The Japanese were losing the war in the Pacific and there was the danger that their country would be invaded and their Emperor killed. It was a situation made for the Samurai warrior whose spirit still lived on in wartime Japan. Youths all over the country volunteered to sacrifice their lives to keep the barbarian away from its shores and considered it an honour to die for the Emperor.

In August 1944 Pilot Officer Ternyoshi Ishibashi, who was undergoing training to be a pilot in the imperial air force, filled in a form which invited officers to volunteer for special assignment. 'They told us that this was a special mission and that the weapon we would be trained for was secret. They also told us that from the time we joined the special unit our lives would not last more than two to two and a half months. We were instructed that when we filled in the form we should make an indication of how strongly we wished to volunteer. If we had no wish, we should make no mark on the paper. If our wish was medium in strength we should make one mark. And if our wish was strong we should make two marks. Some men wrote their wish in blood. I was eighteen years old at the time.'

When Pilot Officer Ishibashi first saw the machine he had volunteered to pilot he was struck by its shape. 'It was a great, big, black thing like an iron coffin. I wasn't scared. In fact I was very keen. I saw no reason to withdraw from my commitment to the project.'

The *Kaiten*, which Ishibashi had seen for the first time that morning, was in effect a huge Japanese Type 93 torpedo known as the Long Lance, modified and lengthened to include space for a pilot at its centre-point. The Long Lance had a 24 inch (60 cm) diameter and was a formidable weapon. It out-ranged the torpedoes of other major navies, carried almost twice the explosive charge of the US Mk 15 and was propelled by a mixture of oxygen and fuel that left no bubble trace which might have given away the position of the submarine that fired it. However, once the United States carrier task forces had established air

OPPOSITE ABOVE A *Kaiten* being launched from the stern of the Japanese cruiser *Kitakami*.

RIGHT 'In Japanese tradition, a leader was always allowed to take his place in the actual attack.' Lieutenant-Commander Mitsuma Itakura was bitterly disappointed when refused a chance to die in a *Kaiten* after training suicide pilots.

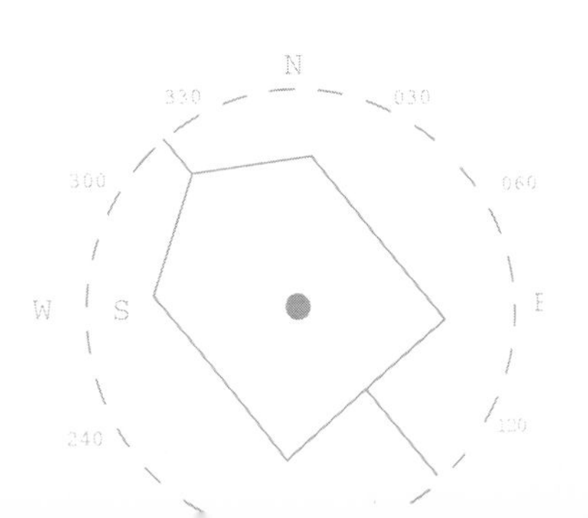
N
330
030
300
060
W
S
E
240
120

supremacy in the Pacific, there were few opportunities for Japanese surface torpedo-boats to engage in battle at a range that could have made the Long Lance a threat as a conventional torpedo. In June 1944 the naval high command began to consider its potential as a suicide weapon.

Senior Flight Officer Yoshiteru Kubo was being trained as an airforce pilot when he realized that there would be a shortage of aircraft for him to fly. When he first heard about the secret project he, like Ishibashi, was told he could make a mark on the paper if he wished to join the project. However, in the atmosphere of the time he knew his superiors were telling him to join rather than asking for volunteers. 'I made a double mark,' he recalled. 'Why? Because Japan was losing the war and in order to serve the needs of the country it was not necessary to stick to aircraft. Anyway, the whole of Japan was geared to fight fiercely against the enemy. If anyone had returned the form without any marks it would have been proof that he was a coward. The whole mood of the nation was mesmerized towards fighting against the enemy.

'They took us to the Kure base to see the *Kaiten* in early September 1944. The blinds were drawn in the train because we were not allowed to know where we were going. It was a great shock when the *Kaiten* was pointed out to me for the first time. At the time of the application they had explained the project was high risk. I thought the risk was because the machine was new. But what I saw there that morning was 100 per cent death. I had imagined that it would be something similar to an aircraft. It was a complete shock. Theoretically it would have been possible to withdraw but in reality I would have been shot. I had no choice but to go on with the project.'

Lieutenant-Commander Mitsuma Itakura, who had been commander of a conventional submarine before he was appointed captain of the training base for *Kaiten* pilots in August 1944, felt cheated. In Japanese naval tradition, a leader was always allowed to take his place in the actual attack. However, in the *Kaiten* programme, Lieutenant-Commander Itakura was not allowed to put his own life on the line. 'I was really determined to pilot one of the *Kaiten* the first time they went into action,' he said, 'but they wouldn't give me permission. The Samurai spirit is the fundamental base of Japanese naval tradition but I was denied my part in that tradition. Because of the psychological dilemma of wanting to go but not being allowed to go, the pressure was getting to me more and more until I became ill and lost weight. The doctor gave me three months to live. Eventually, headquarters agreed that if I was going to die anyway, they'd let me go. Once I got better, however, they suspended permission for me to go.'

Those who were chosen as pilots found their training erratic. Acting Sub-Lieutenant Naoji Kohzu remembered it as being once every three days. And the first half of the day with the *Kaiten* was spent on preparation, followed by only one hour on the craft itself. The following two days consisted of back-up duties with very intensive maintenance courses. 'But a pilot does not do maintenance. There was a definite lack of *Kaiten* available for training and also a severe lack of petrol. At the time everything we heard about the result of *Kaiten* operations was great. There was no talk of failure. We believed that every time a new group of pilots went out with their *Kaiten* they were successful. There were different reactions when colleagues left for a mission. For the military surgeon and the base commander there was pain, but the rest of the pilots felt detached because they knew it might be them tomorrow.'

In the first phase of *Kaiten* operations the naval high command decided to attack ships at anchor in ports. After two missions the Japanese claimed that for twenty-two *Kaiten* lost in action, the Americans had lost four aircraft carriers, three battleships, ten transports and one tanker. The truth was that the entire *Kaiten* effort had definitely sunk only one oil tanker, possibly accounted for an infantry landing craft and had damaged two transports. Fifty or so Americans and more than 200 Japanese had been killed. For the Japanese admirals, who knew the truth, the cost of attacking American ships in harbour was too great.

Only at the last stage of training did Flight Officer Kubo feel he had acquired enough confidence to pilot a *Kaiten*. 'I had twenty training-runs before my first mission. But we were careful when using the *Kaiten* for training. It was far wiser to keep your life for its ultimate purpose and that was the actual attack. Any death during training was a waste. Moreover, if anyone made a mistake in training they were banned from the *Kaiten* programme for a month or even sent away for good. There was some conflict among the pilots about who should be chosen to go on missions. It was a life lived under tremendous pressure. You suffered from your own mistakes but mechanical faults could also ruin a performance and the *Kaiten* was a difficult machine. The biggest problem was that there was no means of communication. So once you were inside your machine and the person outside had locked you in, there was no way you could say anything. And once you were inside, the air supply would run out in a maximum of between four and five hours. There was a tremendous feeling of isolation. You felt like you were inside a tin drum. There was only a short time when you were able to look through the periscope before you submerged. In the main, you could only see dials and darkness. Of course, in training, you know that if things go well you will return, but once on a mission you know you won't. It's a feeling impossible to describe.'

In the last year of the war in the Pacific fewer than ten *Kaiten* out of 150 found any kind of target. Eighty pilots had been killed in action and fifteen more in training accidents by the time Toshitaro Tsakada was assigned to submarine

Ready to die for their Emperor. Senior Flight Officer Yoshiteru Kubo (left) and a friend in 1944 – volunteers for the *Kaiten* programme.

I-363 as a cook. He also doubled up as a starting assistant for one of the five *Kaitens* the I-363 was carrying. On board, the pilots were referred to as 'life gods'. Although they had sacrificed themselves for their living god, the Emperor, they would only achieve the status of demi-gods. Tsakada's pilot was Flight Officer Kubo.

'On its journey to the attack area, the I-363 steered a course between Okinawa and Saipan, where it stopped its engines and, at a depth of 40–50 metres, waited for the sounds of enemy ships,' Toshitaro Tsakada remembered. 'It was around 7.40 a.m. when there was a shout that enemy ships could be heard. The captain ordered the *Kaiten* pilots to get ready and gave directions, angles and positions.

'The *Kaiten* pilots assembled together in front of their leader wearing white headbands and daggers in their waistbands. The captain ordered them to board their *Kaiten*. I took five bottles of lemonade from the fridge and gave them to the pilots as farewell drinks. Flight Officer Kubo was only nineteen years old. It was terribly painful to send such young and lively boys out on a mission of death. They were talking and laughing together until the last moment and yet they were going on a mission of no return.

'Flight Officer Kubo drank the lemonade and said to me, "I am ready. I am going, Tsakada San", and disappeared in the passage to the *Kaiten*. I was aware that the only ties between our submarine and the *Kaiten* were the telephone system and the iron harness which held the *Kaiten*.

'Before the *Kaiten* took off, its tank was supplied with water. I could hear the sound of water running into the tank and at that moment I felt my mind going blank. It was impossible to think about anything except willing the pilot to get to his target by all means. And I was thinking "Go for the biggest target" ... and ... "Don't waste your life on mechanical failure". Under such circumstances it was the only thing to hope for.

'I was holding the release handle of the harness and waiting for the captain's order of "Start". But the order wasn't given and the captain told the crew that the *Kaiten* plan was cancelled.

'When I saw Flight Officer Kubo coming down the passage I felt so relieved that my knees were trembling and I said to him, "How nice that you are back". He answered immediately: "Tsakada San, I am determined to take another mission. Can you give me the lemonade again?" And I said to him, "No, you shall never have one again."'

And he never did. Three weeks later Nagasaki was devastated by the second American atomic bomb and Japan finally surrendered.

CHAPTER 8

TO THE BOTTOM OF THE SEA

On 3 November 1948 Auguste Piccard and his son Jacques were at last inside their bathyscaphe and descending at a steady pace into the unknown off the island of Bao-Vista, one of the Cape Verde islands off the west coast of Africa. They had let the *Trieste* go straight towards the sea-bed. Afraid to throw out too much ballast, Piccard recalled that they did not throw out enough. Three hundred, 400, 500 fathoms (up to 915 m). The *Trieste* continued to drop silently into the depths.

'The projector was turned on and suddenly a circular surface appeared in the cone of light. My son, who was at the porthole, called: "Steady on!" like an aeronaut who expects a rough landing in his balloon. We were already on the bottom: we touched so gently that we were not aware of it. Five hundred and ninety-four fathoms.'

When they turned on their interior lighting they saw that a sandy mass was obstructing the window. Their chamber was sunk up to its portholes in soft mud. In the 'balloon' immediately above them 28 000 gallons (127 260 litres) of refined petrol had been compressed into a fraction of its normal volume. The ballast that kept the bathyscaphe on the bottom consisted of tons of iron pellets held inside one single container – the other had been put out of action before the descent – by an electromagnetic field. Once Piccard cut the electric current to the container, the pellets would run out and the craft would begin to 'fall upwards' towards the surface, becoming increasingly lighter as the petrol expanded in the reducing pressure.

'At the end of a quarter of an hour, thinking it useless to prolong our sojourn at the bottom, we decided to go up. The machine had to be lightened. The opening of one of the ballast tanks was blocked up by a plug; the other was free and allowed 4 tons of iron pellets to be thrown overboard, that is to say more than is necessary to compensate the overload that we had on touching bottom, and to drag the cabin

Piccard, father and son, on the bathyscaphe *Trieste* three years before the US navy sent it to the deepest point on Earth in the Pacific Ocean.

out of the mud. Jacques turned the switch and, in theory, the ballast should have flowed away, but it was impossible in this mud to make sure of it. The silence was total: a real silence of the tomb.

'However, the situation was in no way alarming: a single tank only was available; the iron pellets thus could not flow out faster than a rate of 110 pounds a minute. And even after this tank was emptied, we could still throw the other overboard – in other words, 8800 pounds of supplementary iron pellets and 4400 pounds more, the weight of two ballast tanks when empty. Suddenly the bathyscaphe leant forward and the mud ran along before the porthole. I rushed to it in the hope of perceiving the bottom at last. But, in dragging itself out of the mud the cabin stirred it up; a cloud formed and, when it had cleared away, the bottom was already out of sight.'

As they began to rise and the pressure began to decrease, the petrol in the tanks, with a specific gravity only one-third that of water, began to expand making the bathyscaphe more buoyant and increasing the speed of ascent until it reached 3⅓ ft (1 m) per second. The 'balloon' section of *Trieste*, through which sea-water had continued to flow freely throughout the descent, slowly hauled the weight of the spherical cabin and its passengers to the surface. Both men watched as, in the glimmer of their projected light, innumerable dots showed themselves – particles of mud outlined pale against the black background.

'We were still in darkness. But for the instruments we could still have believed ourselves at the bottom. It is a thrilling moment when the first gleams filter through the portholes! Little by little the illumination grows. From then on there were no more phosphorescent animals. Soon it was light enough for us to recognize objects in the cabin, with all the lights out. The daylight increased and the portholes were resplendent with a bluish light. The cabin began to sway, a slight rocking: we had reached the surface.'

Auguste Piccard and his *Trieste* were to go even deeper. Five years later he piloted the craft down to a depth of 1732 fathoms (3160 m). But seven years after that his son was to take the craft down to the very bottom of the ocean itself.

In 1958 a French navy bathyscaphe – a modification of the original experimental craft Auguste Piccard had developed with naval experts – made nine dives off the coast of Japan during which the pilot, Commander Georges Houot, took the craft down to 10900 ft (3320 m). But Jacques Piccard, now working on new expeditions for the *Trieste* with the United States navy off San Diego, California, was already planning the greatest test for his father's design.

At the southern end of the Mariana trench, 200 miles (320 km) south-west of Guam in the Pacific, the British oceanographic ship HMS *Challenger II* had

The bathyscaphe *Trieste*. The cylindrical 'balloon' attached to the pressurized cabin which protected Auguste Piccard and his son Jacques on their first dives.

discovered a feature where the sea-bed was more than 36 000 ft (1100 m) deep. Inevitably, it was named the Challenger Deep.

Jacques Piccard first heard of it in the mid-1950s in Lausanne. Some time later, Robert Dietz, a United States naval oceanographer with whom he was working, explained that the US navy could open the door for an eventual assault on the Deep. Dietz explained that only the navy could easily support such an operation since they had the facilities, support ships and a naval base near at hand. 'It's true that we can expect a few oceanographers to argue that it is a diversion from more pressing scientific dives,' Dietz explained, 'but I'm sure that most of us will agree: if the capability exists, it must be done.' At that time Piccard was not convinced.

In Washington, eighteen months later, the two men discussed the idea again and this time agreed to make the operation their goal.

'Now that we knew of the Challenger Deep it could no longer be ignored,' Piccard recalled later. 'Until man placed himself on the bottom of the deepest depression on Earth he would not be satisfied. There is a driving force in all of us which cannot stop if there is yet one step beyond. What scientist, dredging up dead husks of life from the abyss, does not yearn to know the whole truth? The bathyscaphe was designed to take that step beyond. Once she touched down in the Challenger Deep there would be no place on Earth, from the highest mountains to the frigid poles, that still thwarted man's entry. It would be the last great geographic conquest. Such an achievement focusing attention on this almost neglected realm would doubtless accelerate the building of deep research submersibles everywhere.'

Piccard commissioned the Krupp foundry at Essen to forge a new sphere to withstand the 200 000 tons of pressure such a chamber would be subjected to at 36 000 feet (11 000 m). The petrol-carrying capacity of the float section was increased by 6000 gallons (27 300 litres) to cater for the buoyancy deficiency caused by the heavier bathyscaphe. The US navy gave the project its blessing and on 5 October 1959 the *Trieste* left San Diego for Guam aboard the SS *Santa Mariana* heading for the last great physical frontier at the bottom of the sea that was the transport vessel's namesake.

On the morning chosen for the dive, Saturday, 23 January 1960, Jacques Piccard was a worried man. When he boarded the bathyscaphe she was being broached by high seas and the deck was a mess. The surface telephone, intended for communication once he and his companion Lieutenant Don Walsh had been sealed inside, had been washed away. The tachometer which measured the rate of speed of descent was badly damaged and inoperative. The vertical current meter was dangling by a few wires. Piccard went down into the sphere. The electromagnetic circuits were in order. By then it was 08.00 hours. He had intended to set off at 07.00 hours. If he did not dive soon there would not be time to complete the 14 mile (22 km) round trip before nightfall.

'According to my calculations we couldn't dive later than 09.00 hours if we were to maintain a safe time margin,' he recalled. 'It is all very well for a man seeking adventure to take chances. I wasn't looking for adventure. I wanted a successful and uneventful operation. I wanted to leave nothing to chance. I made the decision. We would dive.'

A few minutes later water flooded into the antechamber through which the two men had entered the sphere. Without the tachometer Piccard had to watch the pressure gauge to see exactly when the descent began. 'I wanted to log that instant. My eyes were on my watch. Suddenly, at 08.23 hours, the rocking ceased, the sphere became calm, I glanced at the depth gauge. The needle was quivering. We were on our way down. I looked over at Walsh. We both sighed in relief.'

At 340 ft (105 m), 370 ft (110 m), 420 ft (130 m) and again at 515 ft (155 m) the descent halted and the *Trieste* was bounced upwards several yards. At each of these depths the bathyscaphe encountered stubborn resistance where the temperature of the sea changed from surface warm to deeper cold. Never, in all Piccard's previous sixty-five dives, had he encountered so many strong thermal barriers.

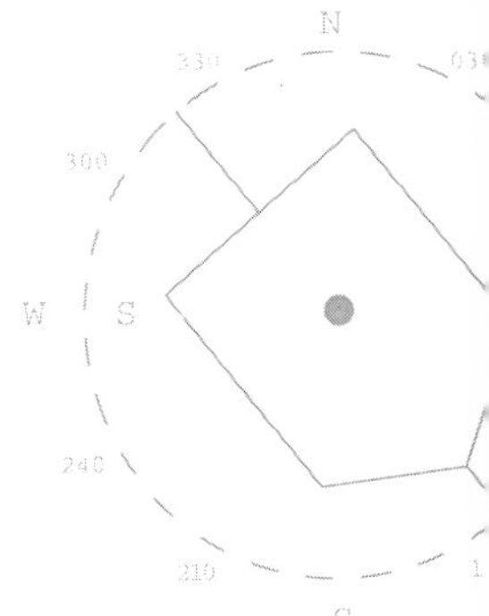

Beyond 800 ft (245 m) they began to drop quickly. At more than 3 ft (1 m) per second they were falling at the speed of an average lift. It was close to the craft's terminal velocity. The plan was to descend the first 26 000 ft (8000 m) at this speed, reducing to 2 ft (0.6 m) per second thereafter and finally to only 1 ft (0.3 m) per second while searching for the bottom. At such a slow rate of descent Piccard knew he had time to discharge ballast to break their speed for landing. 'The charts had warned me that the bottom of the cleft into which we were plunging was a scant one mile in width. Oceanographers have little knowledge of the velocity of abyssal drift. It was easily possible that we might collide with a wall of the trench – a chilling thought. I had to be extremely cautious.

'At 2400 feet we had entered the abyssal zone – the timeless world of eternal darkness. The chill was now penetrating the sphere. Both Walsh and I had been thoroughly soaked while preparing for the dive. Now it was time to change into dry clothing – no simple task in the restricted space of our cell only 3 feet across between the instruments and less than 6 feet high.'

A leak which had allowed water to dribble in at 4200 ft (1280 m) had stopped before they reached 6000 ft (1800 m) but another started when they were near 18 000 ft (5500 m). This one was familiar to Piccard from a previous dive. On that occasion it had sealed itself within a short time. It did so again this time. The craft continued to plunge into the darkness.

'Black water rushed upwards past us. Successively we overpassed the record depths that we had reached in preceding weeks. At 20 000 feet, we were at the maximum depth of the normal Pacific sea-floor. We were dropping into the open maw of the Mariana trench leaving the abyssal zone of the ocean and entering the hadal regions.'

By 11.30 a.m. they were at 27 000 ft (8200 m). Piccard had already dropped 6 tons of ballast to control the speed of descent to not more than 3 ft (1 m) per second. Now he let go more ballast to slow their speed down to only 2 ft

(0.6 m) per second as they came within 10 000 ft (3000 m) of the bottom. Fourteen minutes later they were the equivalent height of Mount Everest below the surface of the sea.

'In the light cone, the water was crystal clear, no "sea snow" and not the slightest trace of plankton. This was a vast emptiness beyond all comprehension. There was, perhaps, a mile of water still beneath us, but the possibility of collision with the trench wall was still on my mind. I pushed the ballast button, slowing us down to 2 feet per second; then, to 1 foot per second, as decided before the dive.'

By mid-day *Trieste* was 31 000 ft (9500 m) down.

'I flipped on the echo sounder and sought for an echo to record on its 600-foot scale. No echo returned; the bottom, presumably, was still beyond 100 fathoms. Trying moments were ahead. We were venturing beyond the tested capabilities of *Trieste*. On paper she could descend safely to 10 miles and the sphere alone much more. I had confidence in those calculations. She was a complex of nuts and bolts, metal, plastic and wire. But a dead thing? No. To me she was a living creature with a will to resist the seizing pressure. Above me, in the float, icy water was streaming in as the gasoline contracted, making the craft ever heavier and heavier. It was as if this icy water were coursing through my own veins.

'Thirty-four thousand feet – no bottom ... 35 000 feet, only water and more water ... 36 000 feet, descending smoothly at 60 feet per minute. Now we were at the supposed depth of the Challenger Deep. Had we found a new hole or was our depth gauge in error? Then a wry thought – perhaps we'd missed the bottom!

'12.56 a.m., Walsh's eyes were glued to the echo sounder. I was watching alternately through the port and at the fathometer. Suddenly, we saw black echoes on the graph. "There it is, Jacques! It looks like we have found it!" Yes, we had finally found it; just 42 fathoms further down.

'While I peered through the port preparing to touch down, Walsh called off the soundings. "Thirty-six fathoms, echo coming in weakly – 32 – 28 – 25 – 24 – now we are getting a nice trace. Twenty-two fathoms – still going down – yes, this is it! Twenty – 18 – 15 – 10 makes a nice trace now. Going right down. Six fathoms – we're slowing up, very slowly, we may come to a stop. You say you saw a small animal, possibly a red shrimp about 1 inch long? Wonderful, wonderful! Three fathoms – you can see the bottom through the port? Good – we've made it!"'

To Piccard the bottom appeared light and clear, a waste of snuff-coloured ooze.

Indifferent to the nearly 200 000 tons of pressure clamped on her metal sphere, *Trieste* balanced herself delicately on the few pounds of guide-rope that lay on the bottom, making token claim, in the name of science and humanity, to the ultimate depths in all our oceans – the Challenger Deep.'

It was 13.06 p.m. The journey had taken 17 minutes less than 5 hours. The temperature of the water was an icy 2.4°C (36.5°F). It had warmed gradually and continuously from the lowest reading of 1.4°C at about 2000 fathoms (3650 m). The depth gauge read 6300 fathoms – 37 800 ft (11 500 m). The gauge had originally been calibrated for freshwater pressures. When recalibrated for salinity, compressibility, temperature and gravity after the dive, the depth attained was adjusted to 35 800 ft (11 000 m) or 5966 fathoms – a distance just short of 7 land miles (11 km).

'As we were settling at this final fathom, I saw a wonderful thing. Lying on the bottom just beneath us was some type of flatfish, resembling a sole, about 1 foot long and 6 inches across. Even as I saw him, his two round eyes on top of his head spied us – a monster of steel – invading his silent realm. Eyes? Why should he have eyes? Merely to see phosphorescence? The floodlight that bathed him was the first real light ever to enter this hadal realm. Here, in an instant, was the answer that biologists had asked for decades. Could life exist in the greatest depths of the Ocean? It could! And not only that, here apparently was a true, bony teleost fish, not a primitive ray or elasmobranch. Yes, a highly evolved vertebrate, in time's arrow very close to man himself.

'Slowly, extremely slowly, this flatfish swam away. Moving along the bottom, partly in the ooze and partly in the water, he disappeared into his night. Slowly too – perhaps everything is slow at the bottom of the sea – Walsh and I shook hands.'

It proved to be Piccard's last meaningful dive in *Trieste*.

Only one year after his record descent, Jacques Piccard speculated on the future of deep ocean vessels. After spending so much time in the *Trieste* he was proud of its immense achievements. But he knew it was extremely limited as an underwater exploration vehicle. It could go up and it could go down, and it could vary its speed along that vertical axis, but the time had come to build in manoeuvrability. Piccard knew where underwater technology was heading.

'Deep ships of the near future built with buoyant hulls will be radically different both from the *Trieste* and conventional submarines. Speed is not so important, for they must move cautiously; the visibility even in the crystal clear bottom water is only 200 feet. Scanning sonars will provide information on obstructions further out. To carry out their scientific mission they will have to be equipped with portholes, prehensile arms, lifting hooks, closed circuit TV, sonar sensors and numerous other scientific devices.'

Exactly six years from the moment *Trieste* had touched the bottom of the sea, every concrete vision of future scientific technology that Jacques Piccard had dreamed of was suddenly needed by the US navy. On the sixth anniversary of his incredible dive the world of manned submersibles was invited to prove that it really had the capability to operate effectively under water – but at a depth and in conditions never imagined before.

It all began when a Spanish fisherman tending his nets off the Mediterranean coast of Spain heard a huge explosion and saw a long object 'like a half-man' attached to a parachute splash down into the sea a short distance from his boat.

The date was 17 January 1966, the day when the tiny fishing village of Palomares, and the area of sea-bed 5 miles (8 km) offshore which fell away to depths of 3000 ft (900 m) below the fishermen's nets, became the centre of one of the most detailed and intensive military investigations the world has ever seen. It was the day when the United States of America lost a hydrogen bomb. Capable of astounding destructive power, it was one of the most vital secrets in the American armoury. And the airforce had no idea where it had gone.

That morning a United States airforce B-52 was returning from a tour of duty in the eastern Mediterranean. At 10.22 a.m. at a height of more than 30 000 ft (9000 m) above the Almanzora river a few miles inland from the coast of Spain it collided with the KC-135 tanker that was about to refuel it for its journey back to North Carolina.

In the resultant explosion both aircraft broke up into hundreds of pieces of metallic debris which showered down on to the Spanish coast. The prevailing wind from the north-west caused the fragments to fall on Palomares, a village so insignificant it did not appear on any of the maps of the area.

Incredibly, not one person in the village was injured as the B-52's tail landed in the dried-up river-bed next to the beach, its landing gear screamed down next to the boys' school and metallic pieces of all shapes and sizes rained on to the village. Seven out of eleven American airmen died in the crash.

The aircraft had been carrying four hydrogen bombs and by noon one villager had kicked one of them while trying to put out a fire and another had been blown over when the TNT element of a second exploded close by. A third was found a short while later very close to the wreckage of the tail-plane.

The H-bombs had not been armed at the time of the accident and were incapable of producing a nuclear explosion. However, the limited explosions from two of them allowed plutonium to be carried away in black clouds and the whole area of the village effectively became radioactive.

The incident was a disaster for the United States which had signed a secret agreement with the Spanish Government allowing its Strategic Air Command to overfly Spain with nuclear devices. Within days it realized that it would have to remove all contaminated material – including topsoil – agree appropriate compensation for the people of Palomares, limit press speculation about the scale of the disaster and convince the Spanish Government of the unlikelihood of such an accident happening again.

It could all have been dealt with relatively quickly had it not been for the fact that, despite all the military personnel drafted in to search every inch of the impossibly harsh and inaccessible terrain around Palomares, there was no trace of the fourth H-bomb. This led the investigating officers back to the testimony of Captain Francisco Simo' Orts who, from the deck of his boat, had seen the aerial collision directly in front of him and above the coastline.

He had already given evidence that he had been waiting roughly 5 miles (8 km) offshore from Palomares for the raising of his shrimp nets. He had seen six white parachutes floating down from the sky. Two had headed straight for him and hit the water behind his boat – the first 80 ft (25 m) away and the second 260 ft (80 m). Other crews in the vicinity supported his account of the parachutes.

Four days after the crash the first United States naval units began arriving off Palomares. For Rear-Admiral William Selman Guest, the fifty-two-year-old career officer chosen to head Task Force 65 in the hunt for the bomb, the prospects were not encouraging. If Orts and the other witnesses were correct, the fourth H-bomb could be 5 miles (8 km) out to sea where depths ranged between 2000 ft (600 m) and 3000 ft (900 m), the sea-bed was ranged with gullies and canyons, there were strong offshore currents and – from the testimonies of Beebe, Piccard and others – visibility without artificial lights was zero.

However, in 1963, *Trieste*, by then without Piccard and totally under the command of US navy personnel, had been used in attempts to examine the wreck of the nuclear submarine USS *Thresher* which had sunk in 8400 ft (2500 m) of water off the coast of New England. Nuclear warheads for some torpedoes were believed to have been on *Thresher* when she was lost and the efforts made to retrieve the submarine continued until August 1964. During those rescue attempts, all the features Piccard had listed, including underwater television, high-resolution sonar and scanners, had been incorporated in the search vessels.

In setting up the Palomares naval search operation Guest had access to all this experience. Twelve days after the disaster a two-man self-propelled submersible, equipped with motors on either side of its hull and capable of navigation, appeared at the site. Flown directly to Spain from the naval ordnance test station at China Lake in California, *Deep Jeep* was the only deep-diving research vehicle available to the navy at that time.

The *Deep Jeep* was part of a huge search team. Admiral Guest had fifteen ships with 2200 sailors stationed off the tiny Spanish village. The shallow coastal waters were being worked by 130 frogmen. Hard-hat divers worked further down. There were twenty naval officers who were experts in underwater work and seventy-five civilian scientists and specialists. The Decca organization had installed a complex navigational system just outside Palomares which would allow precision fixes above the ocean floor. The oceanographic ship USS *Dutton* surveyed the sea-bottom and produced charts from which a model of the sea-bed was built up.

'You can see that it has deep canyons, but off these canyons there are many smaller canyons and gullies and what it doesn't show is that within these are many ditches which vary in depth from 3 feet to over 100 feet. Another thing it doesn't show is that within a small area such as this there are cliffs that are 100 to 150 feet high, sheer cliffs that drop off at 90 degrees,' Guest explained at one briefing at the start of the operation.

This was the landscape in which the bomb was thought to have come to rest. But exactly where? It was going to be a case of searching every square foot of a target area which was estimated to be about 135 sq. miles (350 sq. km). Guest knew he needed civilian as well as military help. Within a few days of taking charge he had sent for the two submersibles – *Alvin* and *Aluminaut* – that were to prove crucial to the needle in the haystack search Guest had on his hands.

Aluminaut was the craft Piccard had envisaged after his dive to the bottom of the sea in January 1960. She resembled a miniature submarine with an overall length of 50 ft (15 m) and a 10 ft (3 m) beam. Painted red, her manned pressure hull was a 6 inch (15 cm) thick aluminium alloy cylinder 33 ft (10 m) long and 7 ft (2 m) in diameter. The need for a gasoline float had gone – the hull had positive buoyancy – but the aluminium hull was not expected to resist the pressure of the sea at depths below 15 000 ft (4600 m).

The *Alvin*, in contrast, was only 22 ft (7 m) long and 8 ft (2.5 m) wide and was much more suited to the craggy, undulating sea-bottom than the larger, less manoeuvrable *Aluminaut*. She could carry a pilot and two observers and could cruise at 2 knots with a top speed of 4 knots. Her battery power gave her a range of 15–20 miles (24–32 km) and she could stay submerged for up to twenty-four hours. Moreover, in addition to her scanning sonar, ground detector and closed-circuit television she also had a grappling arm.

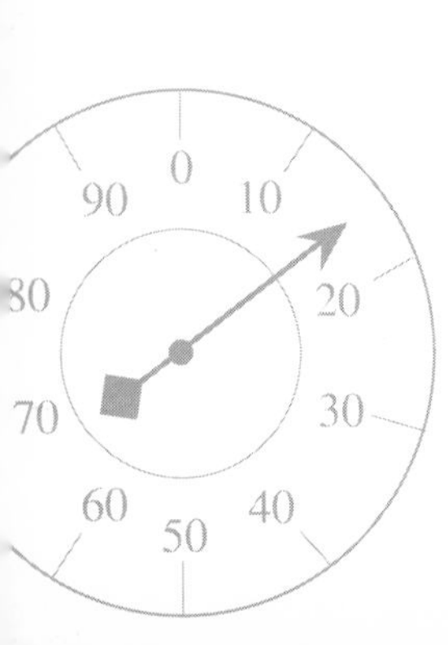

She and her pilots, William O. Rainnie Jr, Valentine P. Wilson and Marvin J. McCamis, were to prove indispensable to the operation. By the time the two civilian submersibles arrived on 11 February Admiral Guest had already lost *Deep Jeep*. It had been sent home with a broken-down electrical system. In

'Like looking for a cigarette lighter in several square miles of wet, rough and rocky ground on a moonless, starless night.' The submersible *Alvin* on the surface after locating the missing H-bomb.

addition, since the start of the operation he had had to put up with appalling weather conditions and very little progress had been made.

When the weather cleared on Monday, 14 February, exactly four weeks after the bomb had been lost, *Alvin* and *Aluminaut* were ordered to dive. Their mission was simple: find the H-bomb. But the *Alvin* pilots were civilians from the Woods Hole Oceanographic Institution in Massachusetts. They had no idea what they were looking for and the navy, for security reasons, was not too keen on giving away state secrets to civilians even when they were searching for one. Eventually an airforce sergeant attached to the naval force, who never revealed his name, took them into a cabin and showed them a photograph of the bomb. They never saw the sergeant again.

'Apart from simply wanting to get the country out of a jam, we had a personal stake in the operation,' Marvin McCamis recalled later. 'This was *Alvin*'s first big job. Our group had proven that the sub could operate to depths of 6000 feet but many critics still considered the entire deep submersible program a waste of money.'

'We began our search about 5 miles from the shore. In this area the floor of the Mediterranean follows a long, gradual slope. But as you move further from the coast, the bottom suddenly drops at a 45 per cent incline. At 2400 feet there is a ridge and the incline steepens to 70 per cent. At about 3000 feet the bottom levels somewhat. Then a short distance away it drops again. Finally at 3600 feet it levels off.

'*Alvin* was very manoeuvrable, much like a helicopter. The main prop at the rear which swung from side to side was for thrust and steering; the two lift props on each side were for moving up, down or around. All were controlled by a single joystick.

'At 1800 feet we had our first look at the bottom. *Alvin*'s powerful mercury-vapour lights gave us 20 to 25 feet of visibility. The bottom was muddy and featureless. Without vegetation it resembled wrinkled old skin.'

Outside that tiny envelope of light the ocean stretched away into unknowable blackness. The task must have been like using a torch capable of illuminating a tiny area, only inches across, to look for a lost cigarette lighter in several square miles of wet, rough and rocky ground on a moonless, starless night.

On 1 March, while taking his turn as surface controller, McCamis tracked *Alvin* far below as it ranged the sea-bed in an area not far from where Captain Orts had reported seeing the parachute and the object that looked like a half-man go into the water. McCamis sought, and was given, permission to break the strict rules laid down by Admiral Guest and move a little outside the area the craft was supposed to cover. Pilots Rainnie and Wilson found nothing promising when they extended their search but they kept taking photographs of the steep slopes and sea-floor gullies they passed over during the long hours of their tour of duty. The craft was operating in 2500 ft (760 m) of water. Towards the end of the dive she crossed what seemed like a man-made feature at a depth of 2600 ft (800 m).

'Wait a minute, I see something,' Rainnie said.

'What?' Wilson asked him.

'I'm not sure, a little to the left, that's it, no, dammit, you went over it. To the right!'

'What?'

'To the right, dammit! That's it. Right on target.'

'What is it?'

Rainnie replied that it had been nothing. Probably some eggshells near a track left by a trawler.

But at the post-dive briefing with the task force leaders, during which the team showed the photographs taken from *Alvin*, the mark on the sea-floor caught McCamis' eye.

'It was a track in the mud that looked as if it could have been made by a skidding object – say a bomb. I hadn't seen the likes of it before and asked for permission to have another look.'

Long dives in *Alvin* over the next three days provided no further information. The submersible was ordered down again on 8 March but this time in a much shallower area. McCamis was annoyed. 'We ought to go back to the fisherman's spot,' he told Rainnie who was navigating at the time. 'That bomb isn't going to float uphill.'

However, the photograph that had caught McCamis' attention had by now also convinced one of the officers on Admiral Guest's flagship. Lieutenant-Commander Brad Mooney spent days trying to persuade the Admiral to allow *Alvin* to dive again in the area indicated by Captain Orts. Guest agreed, with great reluctance that the craft should have 'one more dive' in what was now known as area Alpha One.

On the morning of 12 March *Alvin* dived to try to rediscover the track and the eggshells. McCamis was piloting the craft.

'We were in the fisherman's area. Wilson was at the starboard window. I was in the front and Mark Fox, our mechanic, was at the port window. After about four hours we found the drag mark we'd photographed before and, sure enough, it led downhill at a steep 70 degrees,' McCamis remembered later.

'As I tried to drive down the incline, the stern propeller kicked up a cloud of mud and we lost the track. By now it was already late in the day so we decided to resurface before sunset and let our crew chief get *Alvin* ready for another dive early next morning.

'When Rainnie and I got back, the area looked unrecognizable. *Aluminaut* had been there and the larger and less manoeuvrable sub tore up the bottom. Though visibility was poor we slalomed down a steep slope first going in one direction then in the other. This worked well until we came to what looked like a snowdrift hanging over a cliff. Somehow we made a wrong turn and – thud! – went right into the cliff, unloosing a cloud of mud that thoroughly blocked our visibility. All I could think of was being buried alive. We saw we hadn't triggered an avalanche. When we were finally in the clear we continued the search barely saying a word to each other. It turned out to be another fruitless day.'

Two days later, despite the fact that the navy wanted *Alvin*'s crew to shift to a new area, the civilian team decided to make one last attempt to find the track McCamis had seen on the photograph. They took the craft back down to the area that was now becoming quite familiar. Within a short time McCamis and his colleagues Wilson and Art Bartlett, one of *Alvin*'s electricians, were talking animatedly, oblivious to the fact that their entire conversation was being recorded.

'This looks damned familiar, man,' said Wilson.

'What the hell is that?'

'You've got it right in front of you now. That's the track, that's it, that's the son of a bitch, I'm sure it is, it's the same one. Don't touch down and get it stirred up so we can't see.'

'Yeah, that's it. OK, I'm going in. Better pitch 'em real fast,' McCamis said to Wilson who was the photographer.

'I'm clickin'. I don't know what it is, Mac, but there's two tracks that come down here and converge.'

'OK. OK. Snap pictures.'

'If we hit the son of a bitch, we'll be in great shape. I want to make sure this is it. It sure looks like it.'

'That's it, that's it all right.'

'Echo, this is *Alvin*. We have found the track.'

'I told you I saw the son of a bitch as soon as I saw the bottom.'

'That's right.'

'Goddam, it's steep ... they told me it was a 10 degree slope!'

Alvin was going down an almost vertical slope, where the slightest touch disturbed the sediment, when Rainnie tried to speak to them.

'You just tell him to wait. Tell him we're taking soundings.'

'Echo, we're taking soundings. Wait.'

'We're going down, we're going down fast...'

'OK, I'm going to dive down now. Hold on to your hat.'

Then came disappointment.

'Goddam. I've lost the stinking bottom. Christ. What happened to the bottom?'

'What's the depth?'

'Two four five zero.'

Art Bartlett sat by the bottom viewport. McCamis was looking out of the centre porthole. Wilson was at a side window. They strained to pick up the track once again.

'I see something.'

'The track?'

'Eyup. That's the baby...'

'Ouch!'

'I can see the track! I can still see the track. It's swinging to the left, it's swinging south now. The track is ...'

'I can't come too far. I'm running right into the fucking slope...'

With *Alvin*'s nose close to the sloping side of the cliff, the men could follow the path of the bomb as McCamis began to back *Alvin* down the slope. He was aware that he might become entangled with the parachute still attached to the bomb, but this way all three men could keep the track in sight as it fell away under them. If *Alvin* had driven head first down the slope, the angle of incline would have meant that McCamis would have had to hold *Alvin* in an almost vertical dive to follow the track of the bomb.

'You're coming right into a deep hole...'

'Yeah, I know ... got to do something about this rudder ... How'm I doing now?'

'It's on the right hand side now ... no, no, no, you've got to back straight up to get on it ... back up like a son of a bitch, right rudder down some ... OK, now drive down...'

'It's on that side now.'

'I can't be sure.'

'I tell you I can see it.'

'Coming down ... coming down.'

'About two inches – you're going to hit.'

'That looks like a parachute!' Bartlett shouted. 'A 'chute that's partly billowing.'

'Could be.'

'Open up with the pictures...'

'It's right underneath me.'

'What is it?'

'You know what it looks like, it looks like ... get that fucking squid out of the way. You're spitting all over me! I can't see a thing ... That's it.'

'That is it. I've seen a lot of parachutes and this is a big son of a bitch.'

'What a big bastard.'

'That's where the stinking bomb went! Down this gully.'

'No,' McCamis said. 'It's going to be under the 'chute, I think. Let's go up and take a look.'

'I bet it's gone down the gully.'

'No. Let's look under the 'chute first.'

'Something sure as hell has fallen down into this gully.'

'It's an awful big 'chute, isn't it?'

'It sure is,' came the reply as McCamis edged *Alvin* ever closer to the fabric. All of them were fully aware of the danger should they get entangled.

'I wish I could reach out of this window and pick it up.'

'Take a good look at this one, over the edge. This is where I think the bomb is right here ahead of you.'

'Right here?'

'You know what that is? That's a fin! Mac, that's what a fin looks like on a bomb.'

'Echo. This is *Alvin*. Bill, get as good a position on us as you possibly can. I think we got a big rusty nail down here. We found a parachute and we believe we have a fin of the bomb in sight. It's underneath the parachute.'

Only one word came from the surface, where the news was about to spread like wildfire: 'Roger.'

The *Aluminaut* was immediately ordered down to relieve the *Alvin*.

As the crew of *Alvin* waited in the dark, simultaneously tired and elated, the three men suddenly realized that their craft had moved in the sea-bed currents and that they had lost the bomb.

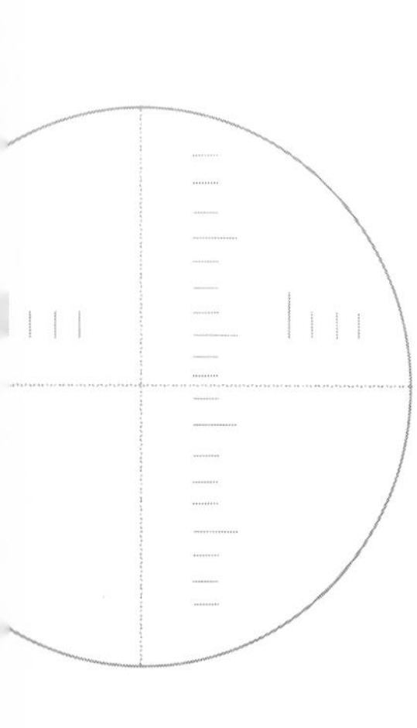

'Jesus Christ...'

'Where in hell is it?'

'I saw it.'

'Yeah but it's just a 'chute.'

'I haven't the slightest idea now where the son of a bitch is.'

'We're too deep.'

'I've got the bastard again. I think...' said McCamis. 'Is that it right in front of us? Shoot it ... I'm going to catch that son of a bitch ... Take pictures hard as you can. Take pictures.'

'I'm moving lots of film through the camera, Mac,' answered Wilson.

'I see something blue...'

'I see something blue, too...'

'It's a blue and gold insignia of some kind...'

'This might be the very fucking nose of the thing. You're going to set down on it in just a second...'

'Look down there and see if you can read that insignia.'

'I'd rather go up and see the other end of it.'

'God. This muddling rudder.' This from McCamis.

'That has got to be the son of a bitch. We're losing it. Oh. Damn it, we're going to murder it. There's nothing I can do with this fucking rudder.'

'OK. Park it. Christ!'

'Ouch!'

'Glad we got out of there.'

'I can see the bomb nose. That's it right at the end. Yes, that's the bomb, no doubt about it.'

'I've got to get out of here before I tangle up in that stuff. Echo? This is *Alvin*. How do you read me?'

'I read you loud but not clear.'

'I think we have enough identification. We'd like to skip clear of this area. There's several straps hanging down loose. There isn't any doubt in our minds about what we see. It's wrapped in the 'chute but part of it shows. The thing is still lodged on a very steep slope...'

'*Alvin*. This is an A-one job. Outstanding.'

McCamis set *Alvin* down in a crevice just below the bomb's huge parachute where the three men waited for eight hours in complete darkness for *Aluminaut* to arrive to confirm the discovery. In turn, *Aluminaut* waited there for twenty-two hours while *Alvin* recharged her batteries and mounted her mechanical arm.

When *Alvin* returned to the site her crew anchored an acoustic signalling device into the parachute with fish-hooks. Over the next seven days of intermittent bad weather, the parachute was anchored by a harpoon. A line to the surface indicated a direct route down to the bomb. By 24 March a 6 ft (1.8 m) square steel frame – an anchor point for three lines with grappling hooks and acoustic sounders – was sitting on the sea-bed about 70 ft (21 m) from the bomb. The plan was to clear the tangle of parachute lines and material around the bomb then hook the lines on the frame into the material and slowly haul the frame, parachute and bomb to the surface.

'We made a special hook for the mechanical arm,' remembered McCamis, 'about the size of a butcher's hook that we hoped could be used to pull away the parachute's shroud lines and untangle the mess in which the bomb was wrapped.

'With our meat hook in *Alvin*'s "hand" we started pulling the shrouds and flaking the 'chute down the slope. Things went well; it looked like we might finally unravel our ball of string. After hooking a shroud, we would back down the slope, making sure the damned 'chute was lying flat so we wouldn't get entangled in it. Then I found that the 'chute wasn't completely out of its compartment on the bomb. So I pointed the hook into the compartment and tugged on a bunch of shrouds. Nothing happened – except that *Alvin* was pulled right on to a 20-megaton hydrogen bomb.

'I tried again, this time fishing out one shroud. I was able to pull it some distance. Slowly, we were getting the job done. All the while, surface was asking us how we were doing. How can you answer when you aren't really sure yourself?'

Back at the site the next day the pilots managed to release the parachute from its compartment and safely stowed it down the slope away from the bomb.

'Now we turned to the frame. Everything was a mess. But we managed to untangle a fine-looking grapple that someone had spent hours making. It was attached to the anchor with a 1-inch nylon line. We managed to attach the grapple right to the top of the parachute shrouds and started pulling shroud by shroud.'

McCamis and his colleagues then pulled the line tight back down slope to the Danforth anchor that was keeping the frame in place. They reported to the surface who agreed that the bomb was ready to be lifted. As a precaution *Alvin* moved to a small hill about 200 yds (180 m) away and waited.

'Then we got word from topside that they'd changed their minds. They were going to drag the entire mess up slope to shallower water before picking up the bomb. I begged them to pick it straight up.' At the back of McCamis' mind was the knowledge that a wet nylon rope, even one that is 1 inch (2.5 cm) thick, will sever like the thinnest strand of cotton when it is in contact with a sharp edge. Lifting the entire rig from where it lay would, in his eyes, be less risky than dragging it across rugged terrain into shallower water. It was yet another argument with Admiral Guest. The *Alvin* team had got their way about going back to search in the area indicated by Captain Orts. On this occasion the Admiral had his way.

The dive had lasted more than ten hours, during which *Alvin*'s three-man crew, working in cramped surroundings, performed feats of unheard of ingenuity and complexity more than 2500 ft (780 m) down on a treacherous sea-bed washed with currents. *Alvin*'s batteries were running low when she returned to the mother ship on the surface.

Shortly after the winches began to wind in the complicated tangle, a sharp edge touched the rope and severed it and the bomb fell back once more into the black depths. The news that it had been dropped came to *Alvin*'s crew as they were waiting to eat. They could not believe their ears.

'The next day, 26 March, Wilson and I returned to where the bomb had been,' McCamis wrote without rancour years later. 'The slope looked as if it had been torn up by bulldozers. We found huge chunks of sand, stone, clay and mud – but no bomb.' It was to take *Alvin*'s crew another seven dives to pick up the trail once again.

'On our thirteenth dive I found an imprint on the sea-bed that seemed to have been caused by the bomb's nose,' wrote McCamis, who described how it even had a bump that matched a dent he had noticed on the bomb's nose the last time he had seen it. 'My suspicion was that the bomb had slid back down the slope and that's where Rainnie and I began to look on the next dive.'

The two men eventually rediscovered the bomb resting in a crevice at the foot of a 70 degree slope 300 ft (90 m) below where they had first found it. They could not allow it to fall any further. If it toppled into one of the gullies that formed the sea-bed around them there might never be a way of recovering it.

The next day Wilson and McCamis went back down to the bomb. They pulled the parachute down slope again and placed another acoustic sounder inside it. This 'pinger' would guide down a new device called *CURV* (Cable Controlled Underwater Recovery Vehicle). It had four ballast tanks, three small motors for propulsion and manoeuvrability, sonar, mercury lamps, a television camera and a large claw for grasping objects and had been developed to recover torpedoes. It was operated from the surface by a five-man crew aboard its mother ship *Petrel*.

McCamis noted that the grapple they had fixed on the parachute eight dives before was still attached to the shrouds with about 50 ft (15 m) to 75 ft (23 m) of line. *Alvin*'s crew stretched out the parachute as well as they could and put another acoustic device in the parachute hole of the bomb. *CURV* was lowered to attach another grapple but when *Alvin* returned the next day there was no trace of the bomb. After a brief search her crew discovered that it had slipped down another 300 ft (90 m) and was now perched on the very edge of a cliff. There wasn't much time left.

On 6 April *Alvin* stood watching as *CURV* paid another visit to the target. As the team on the surface manipulated the submersible to dig in the final grapple they realized that the power in its motors was almost spent.

On the flagship Admiral Guest, frustrated by the delays, decided that he had had enough. Whatever happened now, he at least knew the position of the bomb. He gave the order to cut the bomb loose and leave it where it was. Lieutenant-Commander Brad Mooney could not believe his ears and argued with the Admiral until four the following morning, pleading for one last chance. He got his way. *CURV*'s engineers were ordered to drive the machine on all its remaining power into the billowing material and the cat's cradle of shrouds and

The Palomares H-bomb safe on board a US warship. Rear-Admiral William S. Guest (right) stands with Major-General Delmar E. Wilson in front of *CURV* – the cable controlled underwater research vehicle by which the bomb was winched to the surface.

tapes that hung from it. They did so and the versatile machine became totally enmeshed in the cords and straps that still held the bomb. Mooney ordered the topside winch to haul the whole tangle to the surface. The winch began to turn but heavy seas prevented the surface team knowing exactly what was happening. The winchman called out that the winch-drum indicated that the bomb was 400 ft (120 m) off the bottom. There was 2400 ft (700 m) still to go but the bomb was on its way.

More than an hour later, as the lift entered its final stages, McCamis and Wilson, who had done so much to prepare the bomb for its final ascent, had to watch the operation on sonar as they hovered, impotent, a safe distance away from the lifting area. When the bomb was within 200 ft (60 m) of the surface a team of navy divers went down to fix harnesses and make it totally secure for the final stages of its journey to the surface.

Nearly two hours after the H-bomb had left the bottom of the sea, and as the children of Palomares walked to school that morning, the winchman, concentrating fiercely, gently allowed it to come to rest on the deck of the support ship *Mizar*. A few minutes later the news was flashed to the White House and from there to the rest of the waiting world.

ENTER THE ROBOTS

In late August 1985, almost twenty years after the nerve-stretching conditions faced by Admiral Guest in the Mediterranean, Dr Robert E. Ballard sat in a comfortable air-conditioned cabin aboard the oceanographic research vehicle *Knorr* in the middle of the North Atlantic, studying several banks of television monitors.

Somewhere below him was the wreck of the *Titanic* which had sunk on her maiden voyage on the night of 14 April 1912. The ship had struck an iceberg in mid-Atlantic and only 705 of the 2200 people on board had been saved. Ballard had little more than twenty days in which to find the wreck as *Knorr* had to be back in port to leave on another expedition. He was aware of the immensely difficult task that lay ahead. He had already seen his French partners on the expedition spend three weeks covering the area with their sonar equipment. They had found nothing.

From his position at the plotting-table he was able to watch the pictures transmitted from the bottom of the ocean and move freely about on the sea-bed. There was none of the peril and discomfort he associated with *Alvin* which had been used on innumerable scientific expeditions around the world since its return to the USA after Palomares.

At the start of the first real test of his revolutionary underwater technology, Bob Ballard wondered how long it would be before he and his team caught some glimpse of the evidence they were searching for on the sea-bed more than 2 miles (3 km) down. Everything now depended on *Argo*. Whenever the craft was flying and the search was on, all shipboard operations were concentrated in the

A section of the RMS *Titanic*'s superstructure. The public wanted to see colour pictures before they would believe that Ballard really had found the wreck.

RIGHT Bob Ballard after finding the *Titanic*. 'The French had missed it on the first pass they made. That's how close they were.'

BELOW The *Titanic*. For Ballard the more he read about the ship, the more unbelievable it was. 'All of a sudden you have a real human interest in it. Then it starts to become an obsession.'

control room. Every piece of information relevant to the search appeared on the television screens and every function related to the search could be carried out in that room. There were no longer any underwater time limits. And the days when the crews of manned submersibles had to strain their eyes against the gloom half a mile beneath the surface had gone. Ballard was sure that *Argo* and its successors pointed the way to the future of underwater exploration.

Only one question remained: was the equipment at his disposal sophisticated enough to realize a dream that was now twelve years old? He was convinced that if the wreck of the *Titanic* was inside the target area they had identified, *Argo*, the expedition's 'seeing eye' with its five video cameras, had the best chance of finding it. Towed above the ocean bottom at a speed of between one and 2 knots, the craft could change its depth in mid-flight on a computer command from its operator in the control room. *Argo* had two sonar systems: a forward-looking scanner detected any obstacles in its path and a side-scanning sonar investigated the shape and geological structure of the sea-bed.

The video cameras were to bring the underwater world of the deep North Atlantic to Ballard and his team as they sat in the air-conditioned comfort of the *Knorr*.

The twenty years that had elapsed between Admiral Guest's search for the H-bomb off Palomares and Ballard's search for the *Titanic* had seen an incredible advance in the design and capability of what Ballard called his underwater robots. In the world of underwater surveying these craft are known by the acronym ROV: Remotely Operated Vehicle. By the summer of 1985 the black-and-white electronic video technology of Admiral Guest's *CURV* bore about as much comparison to Ballard's *Argo* as the middle ages do to the space age.

At the beginning of the ROV era, especially after *CURV*'s role in recovering the H-bomb, the world of undersea exploration and surveying had been convinced that it was only a matter of time before new ROV designs eliminated the need for manned submersibles and divers. But it was not to be. Up until the late 1970s a series of manned submersibles were vital to the oil industry around the world. They were mainly free-swimming, battery-powered, wholly autonomous vehicles with the capacity to carry between two and five men and were used either for pipeline observation or to ferry divers to different locations like underwater wellheads and drill-support structures.

But by 1976 the scientific world had gone from having no computers to a position where full suites of mainframe databases were freely available. This development alone was enough to spark renewed interest in ROVs. In 1974 military or scientific projects used about twenty vehicles. Four years later, as a result of the computer revolution, there were 100. By 1980 there were 150.

These developments coincided with demands from oil companies for equipment that would guarantee them longer time at the bottom of the ocean, particularly when pipelines were being inspected.

To many people it was the dawn of an era of underwater robotics. Certainly, there was no limit to the weird and wonderful shapes and sizes of ROV submersibles. And they could offer much greater cost-effectiveness from the word go than previous underwater craft, at a time when the oil industry was moving out of the colossal construction phase of the North Sea oil fields and production costs were becoming an increasingly important factor.

By 1983, when Bob Ballard began thinking seriously about his design for *Argo*, 500 unmanned robot vehicles were in use around the world.

When the US navy agreed to fund a test of *Argo* early in 1984, Ballard knew it would be a race against time if he was to begin any search programme by the summer of the following year. He was convinced that three weeks was not long enough to find the *Titanic*, which could be lying anywhere inside a 100 sq. mile (260 sq. km) area of rugged sea bottom in the middle of the North Atlantic. He had to increase the search time and equipment at his disposal.

Within weeks of the navy's decision, Ballard was in Paris talking to men at the head of IFREMER (the French National Institute of Oceanography) with whom he had worked on a previous expedition. He knew the French wanted a chance to show off a new side-scanning sonar they had developed: 'Since their expertise was almost purely in search technology, it became logical that they should go first. Also they wanted to go first. Naturally I had mixed feelings, because I wanted to find the *Titanic* and clearly I was having to agree to a plan that gave the French the best chance of finding it. And I thought, well, you know, what the heck; it's only what's going to happen. So, I agreed that the French should go first.'

Ballard was pragmatic enough to know that without the extra time and technology the French would bring to the project, there was little chance that he would be able to find the *Titanic* on his own. Moreover, he had obtained the agreement of the Woods Hole Oceanographic Institution on the understanding that his campaign was only to test *Argo*. But unlike his French colleagues, he could not admit publicly that he was really after the *Titanic*.

SMARTIE – one of the remote-controlled survey submersibles used to check structures and pipelines in the North Sea.

For Ballard, the wreck was an unscaled Mount Everest: 'Initially it was basically a contest but then as I began to research the *Titanic*, primarily to figure out where it was, the story just got to me. The more you read about the *Titanic*, the more unbelievable it is and all of a sudden you have more than a technical interest, you have a real human interest in it. And then it starts to become an obsession. People had tried to find it and had failed, very good people, very competent people, which made it an even taller mountain,' he recalled. 'And we felt we could do it. We wanted to prove to the world that we could do it; that we had a new exciting technology in the *Argo* system.'

One example of this technology was a spin-off from the Vietnam War. A starlight scope, a passive light intensifier that allows ambient light to be intensified 10 000 times was applied to *Argo*'s video technology and made it possible for Ballard's team to see clearly large areas of the sea bottom that would otherwise have been completely invisible. *Argo* was also able to fly 100 ft (30 m) above the sea-bed – a measure of control no one had achieved before. These two factors proved crucial in his search for the *Titanic*.

The Americans and French agreed they should concentrate on a primary area of 100 sq. miles (260 sq. km) of sea-bed with a further 150 sq. miles (390 sq. km) of additional terrain as a secondary search area. Everyone was convinced that the wreck of the *Titanic* would be found lying somewhere inside the designated grid – but where? The plan was for the French aboard their command vessel *Le Suroit* to search the selected area with their sonar equipment for the first four weeks of the expedition. Then Ballard and *Alvin* would join them and if they were not successful Ballard and his team would work on for a further twelve days before heading back to Woods Hole. It was all the time Ballard had been allowed from *Alvin*'s tight schedule. Ballard knew he would have to miss the first two weeks of the French search and found himself praying that the wreck wouldn't be found before he joined the expedition.

By 5 July the French had begun to survey the site. After two weeks they had found nothing. They had covered a lot of ground but had been hampered by stronger currents than they had anticipated. Ballard's feelings were mixed when he joined the French team aboard *Le Suroit* at the mid-point of their search: 'I was hoping the French would find the *Titanic* and I was hoping they wouldn't. If they failed, that would leave very little time for the American phase of the expedition to both locate the wreck and photograph it. But it was only human for me to want to be the one to find the ship,' Ballard said later.

The weather totally disrupted the French search and, as the first week gave way to the second and the second to the third, the French team reluctantly admitted that if the *Titanic* was to be discovered that season it would be found by Ballard's team.

Ballard understood the problems posed by the system the French had used: 'The problem with sonars is that they see everything and if you are in a very, very flat area and a ship is all there is then they are perfect. But the ocean doesn't come that way. It commonly comes very rugged. Particularly in the area where the *Titanic* went down, there was a lot of glacial debris, a lot of canyons and gulleys and so many targets that you were overwhelmed by the possibilities.'

Ballard decided on a totally different search strategy. 'Eyewitnesses said they saw the *Titanic* break in half. Well, if you take a ship and break it in half, it's like a salt shaker all of a sudden; things are coming out of it, boilers, safes, wine bottles, deck chairs and everything you can think of is now going to begin a journey to the bottom of the ocean. And that is going to be a long journey for light material and a very rapid journey for heavy material. So you have a classic fall-out taking place. Well, you can model that fall-out. We knew the current was from the north, obviously it was bringing the icebergs down. We also knew the speed and direction of the current was 0.7 knots on a bearing of 170 degrees. And so we could model how objects would fall through that current and be distributed very methodically as a function of their density.

'Our model said that there should be a debris trail that ran north–south and it should be about a mile long with the *Titanic* at the northern end of it. And so that is what we started with.'

Ballard instructed his team to fly *Argo* on east–west lines at right angles to the path of the expected debris trail at 1 mile (1.6 km) intervals. 'That way we only looked at 10 per cent of the ocean floor when a sonar system would have been looking at 120 per cent. So we could move through it much quicker and slice our lines. And sure enough, on about our sixth line as I recall, we came in on debris and we knew to turn north immediately and walk it in. And that's how we found the *Titanic*.'

Only then did Ballard realize how close the French had been to finding the wreck: 'It turned out that all along it had been in the very first square they had intended to search. The French had missed it by a few hundred yards on the first pass they made. That's how close they were. The currents had pushed them off their track but if they had gone on the track they wanted to go on they would have found it the first day. Then they worked themselves further away, never to go back. It was luck – my luck, their bad luck. But the reason we found the *Titanic* was that we didn't pull *Argo*. We kept it down and worked it round the clock. And not only was I able to see what was going on, but so could anyone that walked into the control room.

'The terrifying part about the *Titanic* programme was that the public wanted pretty pictures and *Argo* was a television camera system that took fuzzy black-

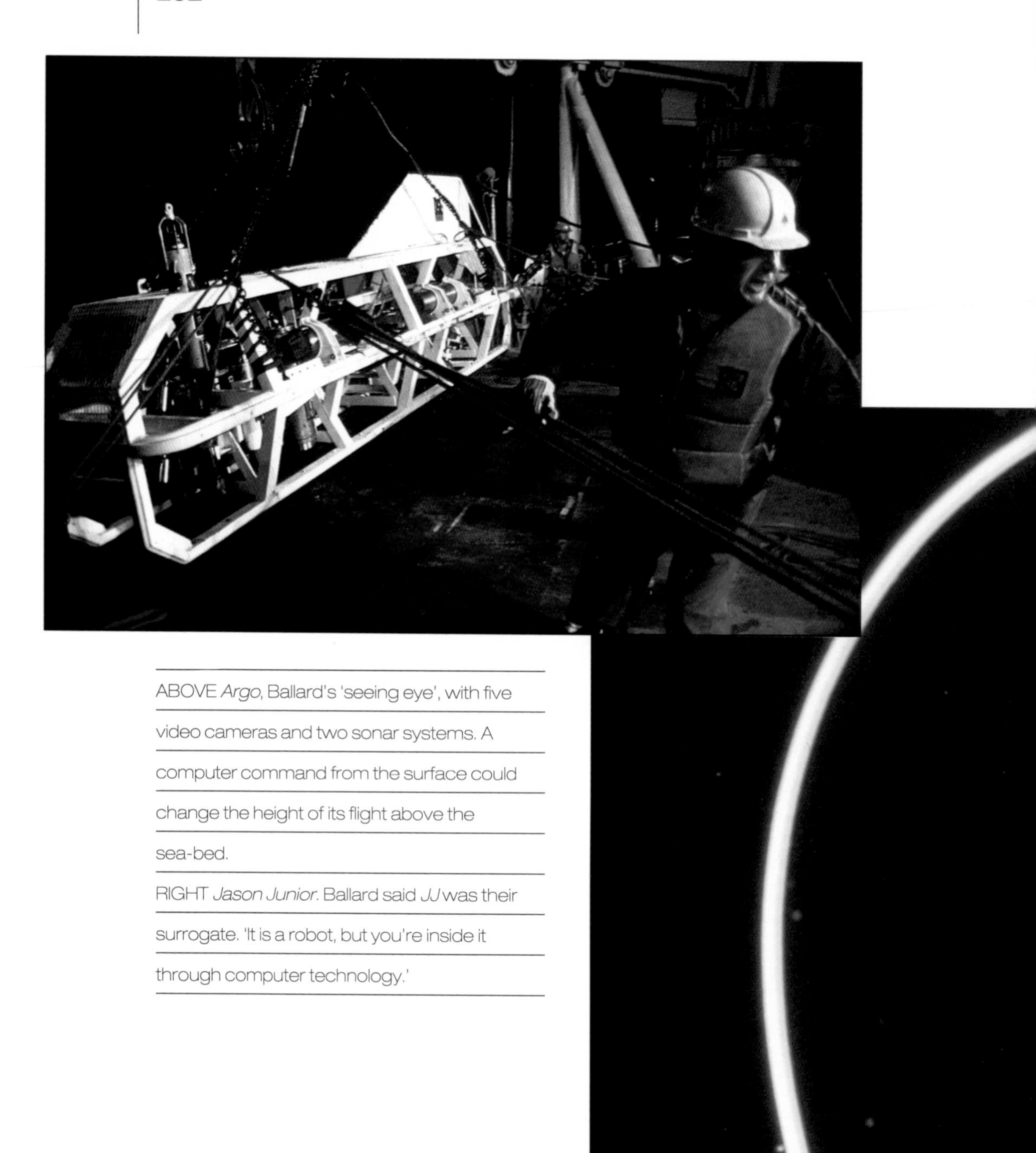

ABOVE *Argo*, Ballard's 'seeing eye', with five video cameras and two sonar systems. A computer command from the surface could change the height of its flight above the sea-bed.

RIGHT *Jason Junior*. Ballard said *JJ* was their surrogate. 'It is a robot, but you're inside it through computer technology.'

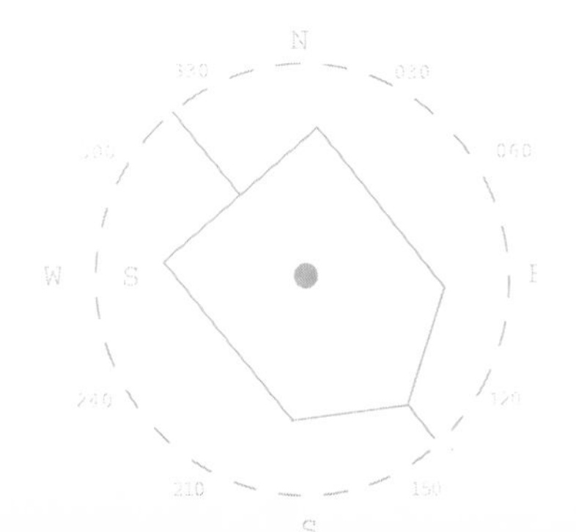

and-white images. And after we said we'd discovered the *Titanic*, people said, well, show us. We sent ashore a black-and-white video and they said: "That's it?" and they said they wanted some colour pictures. So I did the most terrifying thing I've ever done in my life which was to cause one of our towed vehicles to make a blind run along the axis of the *Titanic* without knowing where the camera was in relation to the wreck.' But Ballard got the beautiful pictures the public had clamoured for and the world was at last convinced that he had found the *Titanic*.

In late July the following year, his team returned to the wreck in a new ship, *Atlantis II*, and a refined *Alvin* which could descend the 2½ miles (4 km) to the ocean floor with Ballard on board along with the two-man crew. *Argo*'s place

had been taken by a much smaller remote-controlled vehicle, *Jason Junior*, which everyone referred to as *JJ*. *Jason Junior* was a powered, tele-operated, underwater robot operating under orders from *Alvin*. 'It was our surrogate,' Ballard explained. 'It is a robot, but you're inside it through tele-presence technology. You can order it to be fully automated or control it completely through computer technology. *JJ* was designed in the first place to go inside the *Titanic*. It was extremely compact and small and had very reliable motors because once you get it inside you don't want it to die on you. It permitted us to land on the *Titanic* and then send *JJ* down the grand staircase or over the side, along the promenade deck and to go into very dangerous areas with some degree of certainty that we weren't going to lose everything.'

Ballard will never forget his first experience of descending on to the *Titanic* in *Alvin* and watching the pictures sent back by *JJ*. His goal had always been to go inside the ship. He could have gone down the smoke-stacks but that was not the journey he wanted to make. The *Titanic* had been known for its opulence and that was reflected in the first-class areas of the ship. His research had indicated that on its uppermost deck, between the No. 2 and No. 3 funnels, a huge glass dome had covered the grand staircase that led from the first-class entrance. That had become Ballard's target for *JJ*.

'We landed on the deck and had no idea if it would hold my submarine,' he recalled. 'I figured it had to because it looked pretty sturdy but we had our hands on the emergency release weights in case we started to collapse into the ship. But we didn't. And right in front of us was this abyssal opening, this giant black hole. We watched *Jason Junior* go down that hole and quickly lost visual contact. And then we began to look at what *Jason Junior* was seeing and it was going down and it became a giant elevator shaft. Then we would look and follow the wall down and we'd periodically stop and look around and we were terrified. And we weren't even there, we were up above. We were terrified. And we dropped down, dropped down, and then we saw this open area and I felt we'd gone deep enough, it was about three decks down and I said, "Now leave the elevator shaft and enter one of those rooms". Martin, *JJ*'s pilot, looked at me. So I said, "You're not going to not do it? We've got to do it. We didn't come here to sit in the elevator shaft". So we then started off in this room and there was a pillar we went by and then all of a sudden, out of this gloom, I saw a shiny object suspended in space and, as we panned, our lights would bounce off it and I said, "Go over there". And we came in on this chandelier hanging from the ceiling. I couldn't believe it and we sat there and we just looked at it.'

Ballard dived on the *Titanic* ten times that season recording, examining, photographing and searching. The wreck itself, broken into two sections driven deep into the sea-bed and melting into rust, had been the prize. But the trail of debris on the sea-bed – a pair of shoes, a child's doll, deck chairs, chamber pots and a ship's safe – also stirred the imaginations of the men who saw her. When Ballard left the wreck for the last time in the summer of 1986, he knew he would never go back.

The expedition had been an overwhelming success. With new equipment and new technology his most recent work, on the wreck of the *Lusitania*, has been done not from a ship at sea but from the comfort of his own home where satellite and video technology now allows him total control of every aspect of his expeditions.

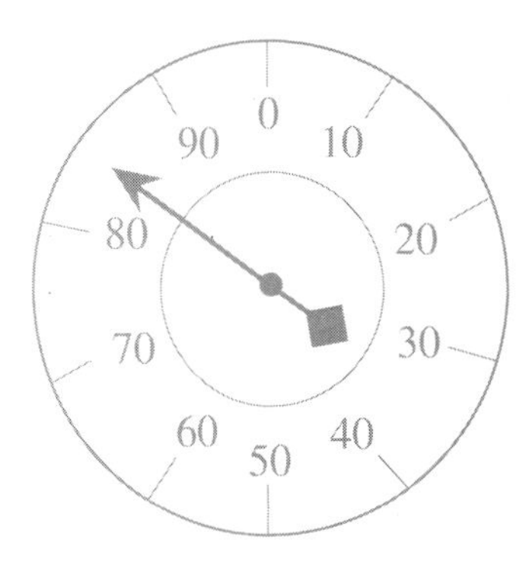

MASTERS OF INNER SPACE

Admiral Hyman George Rickover is universally acknowledged to be the most important and influential individual in the development of the United States' nuclear submarine. He had been so almost since the end of the Second World War and has been credited with organizing the design, launch and commissioning of *Nautilus*, the world's first nuclear submarine, five years before many believed such an achievement was possible. From the very beginning he made himself personally responsible for the development of his country's nuclear fleet. He not only had the last word on the design of the reactors used by the United States navy but also demanded a say in the speed, depth, safety and operating standards of almost every nuclear project from drawing board to slipway. Hyman G. Rickover was a very powerful officer. What was even more important was that Hyman G. Rickover almost always got his way.

But one crucial factor in his professional life was outside his control. Although he could constantly check and approve every aspect of his own nuclear fleet, he could not monitor the actions of a man several thousand miles away whom he had rarely seen. There was only one other admiral as powerful and influential as Rickover in the world of nuclear warships and that was Sergei G. Gorshkov, Commander-in-Chief of the Soviet navy, who had his eyes firmly fixed on his American counterpart and the developing capabilities of the US navy.

By the time the careers of both men came to an end in the 1980s, Gorshkov and Rickover, both blindfolded by state secrecy as to each other's achievements and intentions, had been locked in a game of underwater nuclear chess that had lasted for more than twenty-seven years.

USS *Billfish*, a Sturgeon class attack submarine, surfacing at the North Pole in 1987.

Admiral Sergei Gorshkov when he was Commander-in-Chief, with the power to shape the nuclear future of the Soviet navy.

Sergei Gorshkov, born in 1910, entered the Frunze Higher Naval School in Leningrad in 1927 and spent four years there before being commissioned to serve on surface ships in the Black Sea fleet and the Soviet Pacific flotilla. Seven years later, at the age of twenty-eight, he was given command of a destroyer and within a year he commanded a destroyer brigade in the Black Sea, before being selected to go on a course for senior officers at the Voroshilov Naval

Academy, the Soviet naval war college. His early promotion was not simply the result of his own ability. Stalin purged several thousand serving officers during the late 1930s and Gorshkov inherited one of the commands which became vacant. At the outbreak of war between the Soviet Union and Germany in 1941, he was appointed Rear-Admiral – a position in which he served with distinction in the Black Sea, the Sea of Azov and on the Danube. By 1951 he was Commander-in-Chief of the Black Sea Fleet and four years later, already an Admiral, he was appointed First Deputy Commander-in-Chief of the Soviet navy. One year later, in 1956, at the age of forty-six, he became Commander-in-Chief and a Deputy Minister of Defence. Admiral Gorshkov now had the power to shape the nuclear future of the Soviet navy in order to counter the threat of Western aggression towards Communism and the Soviet motherland. It had been a meteoric rise by any standards – and a complete contrast with the frustrations and set-backs encountered by Admiral Hyman G. Rickover who, also in 1956, already held the potential of the United States' nuclear navy in the palm of his hand.

Hyman G. Rickover entered the United States Naval Academy in June 1918 and graduated 107th in a class of 539 graduates four years later. Like Gorshkov, his first assignment was to destroyers, before he was assigned to serve on battleships. In 1928, after attending naval postgraduate school and Columbia University, where he earned his Master of Science degree in electrical engineering, he volunteered for submarine duty.

The US navy submarine branch he joined was small and the boats were so dirty they were known as pig-boats. They were cramped and smelled and diesel fumes penetrated the clothes and skin of every man aboard. But submarines offered the opportunity of early command to a newly appointed full lieutenant who was seeking experience in electrical propulsion. Between 1929 and 1933 Rickover served in submarine S-48 as engineer and electrical officer and later as executive officer and navigator. By the end of this period he became eligible for command of a submarine but it was never offered him. During his time with S-48, Rickover was never accepted by the officer élite of the Submarine Service.

It has been suggested that the bitterness Rickover showed towards his naval contemporaries throughout his career stemmed from his experiences while serving on S-48. In later years he referred to the United States' naval bases at New London, Charleston, San Diego and Pearl Harbor as 'social centres' and 'clubs' for the submarine community. One retired officer commented: 'Rickover just never fitted in. He couldn't get along with people. He was passed over [for command] . . . because of his abrasive personality'.

After brief command of a minesweeper in China, and two years as an engineering officer in the Philippines, Rickover was assigned to Washington

Admiral Hyman G. Rickover. His ability to undertake assignments of great difficulty far outweighed the personality problems that had antagonized fellow offices throughout his career.

where he spent the war years in charge of developing and buying electrical systems at the Bureau of Ships at navy headquarters.

When the war came to an end, Rickover, now a captain, was an engineering duty officer with twenty-three years' experience but, like thousands of other senior naval officers, he could not see much of a future for himself. The United States' battle fleet was in the process of being reduced to one-tenth its peak wartime strength and Rickover was posted to the West Coast as an inspector-general supervising the mothballing of units of the 19th Fleet.

A few months later, on 29 March 1946, 3000 miles (4800 km) away on the East Coast, Philip H. Abelson, a scientist working with the Carnegie Institution, outlined his ideas for a nuclear reactor to fit into a submarine the Germans had built at the end of the Second World War. Abelson's report, *Atomic Energy Submarine*, contained little information about how the nuclear reactor would be designed, but he proposed the use of a sodium-potassium alloy as the means to transfer heat from the reactor to the steam turbine which would drive the propeller. He concluded his report: 'A technical survey conducted at the Naval Research Laboratory indicates that, with a proper programme, only about two years would be required to put into operation an atomic-powered submarine capable of remaining submerged for months at a time without needing to re-surface or refuel. Its speed would be in the region of 26 to 30 knots [almost one-third as fast again as the submarines the Germans had been building at the end of the war]. In five to ten years a submarine with probably twice that submerged speed could be developed.'

Abelson was not the first American scientist to suggest the use of atomic power to drive either a surface ship or a submarine. However, he and a colleague, Dr Ross Gunn, of the Mechanics and Electricity Division of the US Naval Research Laboratory in Washington, were the first to get enthusiastic endorsements for the idea from senior submariners in the US navy. 'If I live to be a hundred,' recalled Vice-Admiral Charles Lockwood, commander of all submarines in the Pacific during the war, 'I shall never forget that meeting on 28 March 1946, in a large Bureau of Ships conference room, its walls lined with blackboards which, in turn, were covered by diagrams, blueprints, figures and equations which Phil Abelson used to illustrate various points as he read from his document, the first ever submitted anywhere on nuclear-powered subs. It sounded like something out of Jules Verne's *Twenty Thousand Leagues Under the Sea*.'

The Gunn and Abelson initiative was timely. Only two weeks before, James V. Forrestal, Secretary of the Navy, had indicated to Robert Patterson, Secretary of War, that the navy wished to undertake the engineering development of the atomic power that had been used to devastate Hiroshima and Nagasaki. Patterson invited the navy to assign personnel to an atomic project that was being set up at Oak Ridge, Tennessee. Its aim was specifically to prepare a group of naval officers for the time when a sea-going reactor plant would be developed. The best guess at when that time might be was between four and five years.

One of the men who was selected by the navy to undertake parallel studies to develop a nuclear plant for destroyers at the General Electric Atomic Power Laboratory at Schenectady, New York, was Captain Hyman G. Rickover. When the final assignments were announced, however, Rickover had in fact been appointed senior officer of the five-man navy group that was to observe the nuclear reactor project at Oak Ridge. Rear-Admiral Earl W. Mills, who was responsible for the selections, was Deputy-Chief of the Bureau of Ships in the Department of the Navy and had been Rickover's wartime boss. He decided that Rickover's ability to undertake assignments of great difficulty successfully – and the Oak Ridge project would certainly be difficult – far outweighed the personality problems that could, and had, antagonized fellow officers throughout his service career. Admiral Mills was gambling that Rickover, in his inimitable way, would find the ways and means necessary to defeat the bureaucratic problems the navy would inevitably encounter as it moved towards the construction of a nuclear fleet. He had watched Rickover absorb incredibly detailed and intricate technical material and remembered that when Rickover had been head of the electrical section in the Navy Department during the war he had made it the most competent area of command in the whole of the Bureau of Ships.

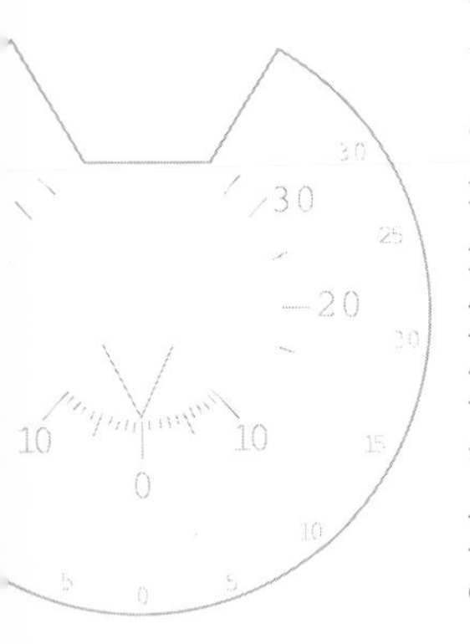

The nuclear assignment to Oak Ridge was to be the making of Captain Hyman G. Rickover. Even before leaving California he began to study nuclear physics, chemistry and mathematics and, once in Washington, went through all the navy files in the Bureau of Ships that related to nuclear matters. One month later he joined the army of scientists, planners and engineers who would be part of the Oak Ridge project. Over the weeks that followed, they discussed and theorized about the practical use of nuclear reactors in naval ships with atomic scientists and with contacts from the huge industrial corporations of Westinghouse, General Electric and Allis Chalmers who were to work on future naval nuclear projects. This wealth of knowledge and experience that was available within and without the military in the United States was to contrast dramatically with the limited knowledge, secretiveness and compartmentalization with which the Soviet Union began to tackle their nuclear submarine project six years later.

In the autumn of 1946 Rickover, already acknowledged by some as the navy's authority on nuclear power, part-authored a report that provoked debate about the possibilities of nuclear propulsion for both submarines and surface craft. It predicted the production of the first nuclear-propelled submarine within five to eight years and a fleet of nuclear-propelled warships within ten to sixteen years – a target which the report warned could only be met if a great deal of engineering work was undertaken over the same time period.

Shortly afterwards Rickover was appointed Admiral Mills' special assistant for nuclear matters. He was at last in a position where he could not only organize the building of a submarine driven by nuclear power, but could also depend on the navy's widespread support for nuclear-powered submarines in defeating the bureaucracy that might have held back the development of the craft. Within months he was also appointed Navy Liaison Officer to the civilian Atomic Energy Commission, an appointment that effectively put him in charge of the navy's civil and military nuclear propulsion programme and in a position to dictate the speed of its advance.

With Admiral Mills' backing and the knowledge that he had been given responsibility for the navy's entire nuclear reactors research programme, Rickover introduced a regime dedicated to the highest standards of workmanship and engineering skill. He worked as hard and as tirelessly as any of his team and constantly checked the progress and efficiency of every aspect of his atomic programme.

In 1949 a nuclear reactor with a heat transference system based on pressurized water was ordered by the navy. It was to be a full-scale prototype based near the desert town of Arco, Idaho. The construction of the submarine in which the nuclear reactor would ultimately be fitted would have to begin in 1952 if the vessel was to be completed by 1955. Rickover kept the project on schedule. He demanded efficiency, argued ferociously with workmen, designers and engineers alike, complained bitterly at the navy's way of doing things and generally upset or checked up on people twenty-four hours a day. The first invitation to build a nuclear submarine was offered to and accepted by the Electric Boat Company of Groton, Massachusetts, which had built John Holland's submarine boats before the turn of the century.

The reactor worked for the first time on 30 March 1953. Two months later it fed power to a turbine and on 25 July, as full power was achieved, Rickover ordered the throttle be left open to simulate a submarine crossing of the Atlantic. The plant had to be throttled back three times at moments of minor concern, but the reactor never stopped powering turbines which, if fitted inside a submarine, would have driven the vessel to Ireland in ninety-six hours.

America's first nuclear submarine, *Nautilus*, was launched on 21 January 1954 and commissioned on 30 September. Her hull design was based on the general lines of the streamlined German Type XXI submarine which dated from 1944.

All earlier submarines had had to surface to get the air needed to operate their diesel engines. But the Germans had designed the Type XXI to operate continually under water by fitting a snorkel device which conducted air to its diesel engines while it ran submerged. The Type XXI did not have the traditional bow that marked most earlier submarines, or a large deck gun or other projections that might cause water resistance. These improvements, along with more powerful batteries, were important factors in its ability to travel submerged at 18 knots when other Second World War submarines were only capable of about half that speed.

However, if nuclear-powered submarines were to increase their speed significantly, they needed to be much more hydrodynamically efficient under water, and this required an entirely different shape of hull such as the teardrop which the Russian designers were to choose for their own nuclear submarines.

But in the heady days of spring 1955, it was *Nautilus* that was the most exciting warship afloat and Rear-Admiral Hyman G. Rickover, the man who had made it all happen, was a national hero.

When Sergei Gorshkov was appointed Commander-in-Chief of the Soviet navy in 1956, he may have known a great deal about Rickover and *Nautilus* but may still have been unaware of his country's own K-3 project. The Soviet nuclear submarine was still being built, in conditions of such secrecy that even the most senior government officials did not know of its existence. However, the project would have been one of the first things confided to Gorshkov along with the assessment that the Soviet Union was perhaps four years behind the Americans in nuclear submarine technology. With Soviet political and naval pride at stake, Khrushchev's order to catch up and surpass the Americans applied to Gorshkov as much as it ever had to Vladimir Peregoudov.

It was six years before K-3 successfully navigated under the ice at the North Pole and allowed the Soviet Union to show the world that they were catching up with the Americans. But those years were crucial to the second part of Khrushchev's instruction. Gorshkov could not afford to wait for K-3's success. In the late 1950s he and the rest of the Soviet navy began a programme of long-term planning that was to have an impact on the military and political world far beyond the end of the next decade.

Admiral Gorshkov decided to adopt a strategy based on a three-strand path

A US navy attack submarine submerged to periscope depth.

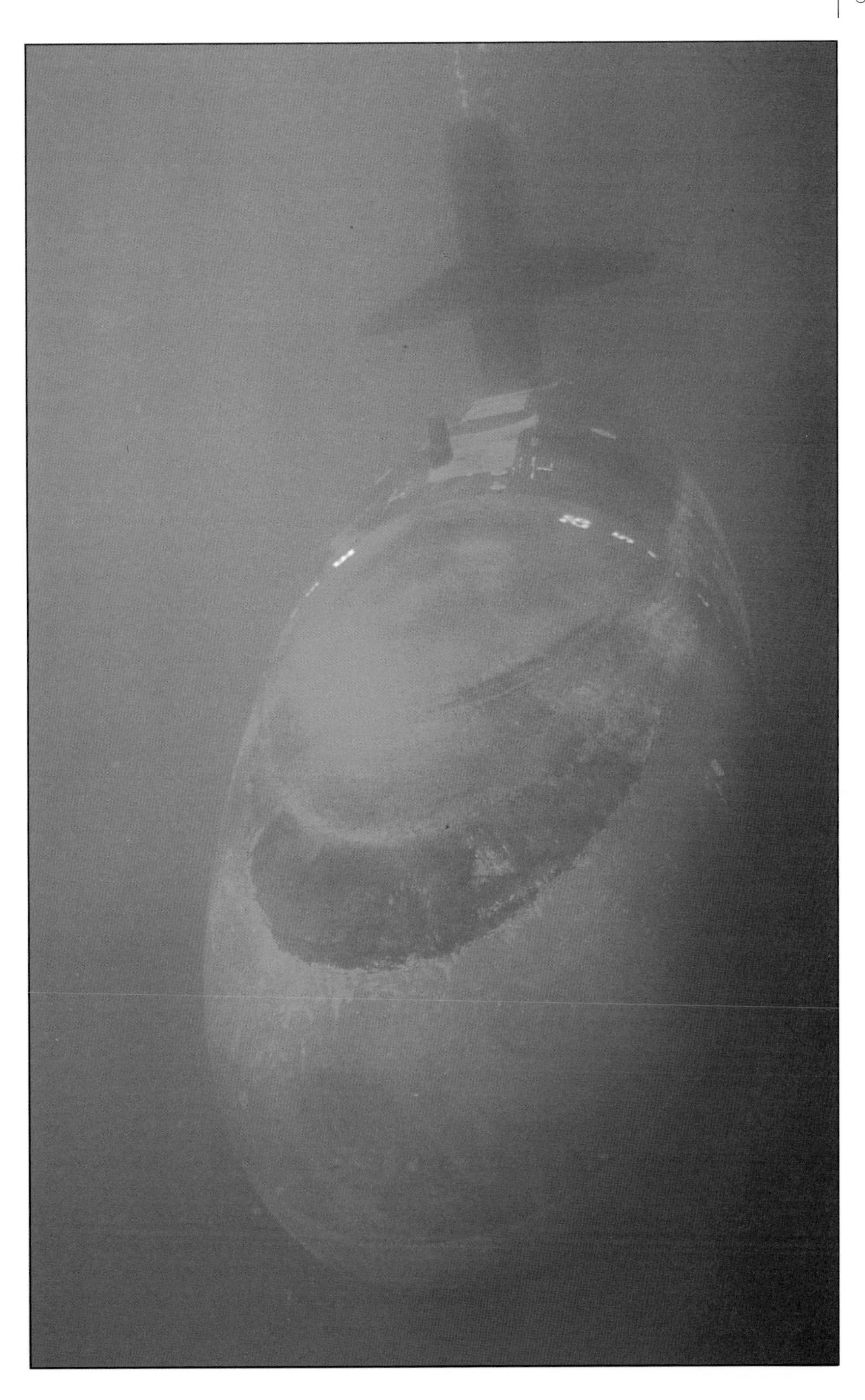

593
593

of Soviet nuclear submarine development. He would build fast attack submarines to search out and destroy enemy surface ships. He would build submarines which, armed with short-range guided missiles, could be used to attack land-based targets; and he would design a series of ballistic-missile submarines which, from far out to sea, could launch rocket attacks on enemy cities far inland. Western intelligence services knew them respectively as the November, Echo and Hotel classes of Soviet submarines. They were destined to become the shining stars of a Soviet fleet that still contained many conventional diesel-electric submarines.

By 1962, when the Cuban crisis came close to causing a nuclear war between the superpowers, Gorshkov had eight nuclear ballistic-missile submarines at his disposal which matched the fleet of Polaris ballistic-missile submarines the Americans had at sea in number, but not in firepower or range. The Soviet Union was determined to catch up with and pass the nuclear capacity of the United States in the shortest possible time and with a submarine fleet based on Peregoudov's teardrop design and a double-reactor power unit, both standard elements of the original K-3, they had the means. However, it was only in 1967, when the US nuclear submarine fleet had fourteen boats in the Thresher and Sturgeon classes of torpedo attack submarines and forty-one Polaris and Poseidon ballistic-missile submarines, that Gorshkov was able to respond. In that year the Soviets reacted to the American build-up by launching the largest ballistic-missile submarine they had yet built – but one which still only matched the size of the earliest Polaris submarines built seven years before. The Americans designated it the Yankee. It displaced 9600 tons when submerged and carried sixteen ballistic missiles, each with a larger warhead than the Polaris but with only half the striking distance. The comparisons were discomforting to Western analysts but not by any means disastrous. The Soviets were catching up but in missile technology the United States still had the edge.

Then, in July 1968, Admiral Rickover told the Joint Congressional Committee on Atomic Energy: 'Last year in testimony before Congress I stated that in my opinion, the Soviet Union would surpass us in their nuclear submarines within five years. I still hold to this view, although I may have underestimated the Soviet advance.'

USS *Thresher*, a nuclear-powered attack submarine, which sank with all hands off the New England coast in 1963.

The Victor III class attack submarine. With its introduction the Soviets again reduced the technological inferiority of their latest submarines compared to those of the United States.

A few months after that statement, Rickover was briefed by US Naval Intelligence that the Soviets had nuclear attack submarines at sea which could travel submerged at a speed comfortably in excess of 30 knots for an indefinite period of time. In one incident the nuclear-powered aircraft-carrier USS *Enterprise* had engaged all eight nuclear reactors which drove her turbines and had approached her top speed of more than 30 knots between San Francisco and Pearl Harbor in the central Pacific without being able to shake off a following Soviet submarine. It was a deeply worrying scenario. An aircraft-carrier outrun by a submarine was unthinkable. If Soviet attack submarines were as fast as a carrier, the entire United States surface fleet was vulnerable. Rickover realized the dangers better than anyone but few moves by which he might redress the balance were left to him.

In the late 1960s Rickover began arguing for government funding for a new Los Angeles class of attack submarine with a more powerful reactor that would give it a top speed of 33 knots. 'The rapidly increasing Soviet threat makes it essential that the United States get the new high-speed class into the fleet as soon as possible,' he told the sub-committee of the House Committee on Appropriations. The congressmen listening to his arguments for the new submarine were disappointedly aware that although the new class would be 5 knots faster than the Sturgeon the most recent US attack submarine, the cost would be much, much higher. It was not something an administration paying out tens of millions of dollars a day for the war in Vietnam wanted to hear.

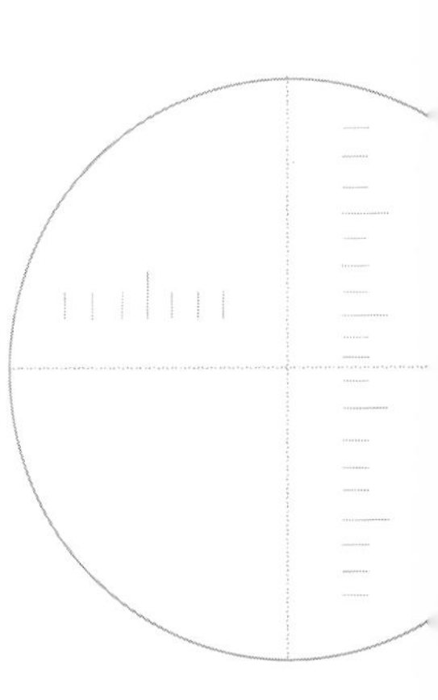

Rickover eventually got his way on the Los Angeles class, but it was to do him little good. Gorshkov's designers were already working on a new attack submarine which would have the speed and acoustic levels to match the Los Angeles and which, in most respects, would be superior to the capabilities of Rickover's new boat.

The Soviet determination was remorseless. A Director of US Naval Intelligence was to admit that, with the introduction of this new class of submarine: '. . . the Soviets have steadily reduced the technological inferiority of their newest submarines. Progress in Soviet submarine quieting and sonar signal processing improvements have reduced the acoustic advantage of Western SSNs [i.e. attack submarines]'.

Western military analysts believed that Gorshkov wanted the new craft, which they had designated the Victor class, to help defend the Soviet motherland from the threat of US missile carrying submarines as well as being able to help defend the Yankee Soviet ballistic-missile submarines from American and British attacks from directly astern.

The Victor, with its streamlined hull, could make more than 32 knots, a speed equivalent to that of the Skipjack class, the fastest United States' attack submarines. It could also launch a 20 mile (32 km) missile with a conventional torpedo as warhead. The pace of Soviet nuclear submarine development could no longer be ignored. It was a situation which worried the navy, Congress and the presidency.

By the middle of the 1970s, with the Los Angeles class still more than a year away from its launch date, and the Soviet Union ahead in both numbers and performance in its attack submarines and with the new Yankee class and a new Charlie class of guided missile submarines already on patrol, Rickover turned on Congress for denying him the funds necessary to improve his own nuclear navy. He declared publicly that the competition between the United States and the Soviet Union to produce the best attack submarines had not been an arms race: 'The Soviets have been running at full speed all by themselves,' he said.

Certainly the Soviet Union's expenditure on its submarine fleet would have been difficult to match, but since the earliest days Rickover had demanded, and been given, complete control over the design and performance of succeeding classes of American attack submarines. He had so manipulated the US navy and the civilian authority, the Atomic Energy Commission, that his control over all contracts for nuclear commissions for the US navy was total. However, the Soviet Union had taken the initiative ever since the Skipjack class of attack submarines in the early 1960s. Soviet submarines had got faster while the Thresher and Sturgeon classes, though quieter and more electronically sophisticated, had got slower. Criticism was aimed at the all-powerful Hyman G. Rickover. But there was no short-term or easy answer.

The problem was that, in trying to increase the speed of his newest submarines, Rickover had been forced to increase the size of the nuclear reactor to develop the greater power needed to push the larger vessel through the water. But the bigger the size of the plant the bigger the submarine had to be to carry it, and this partly nullified the extra power of the reactor. And Rickover, like the Russians at this stage, was wedded to the system whereby nuclear reactors used pressurized water both to cool the reactor and to conduct the heat that turned water into steam to drive the turbines. Twenty years before he had tested a prototype, *Seawolf*, with the potentially more efficient sodium-potassium metal heat-transference system originally proposed by Abelson before rejecting it on safety grounds in favour of the pressurized-water reactor. And once Admiral Rickover had set his mind against something it was not his way to allow it to be reconsidered.

In the early 1970s the Department of Defense, concerned by the inferior performance of their attack submarines, began to reconsider the possibilities of liquid metal-cooled reactors. Indications were that the physical properties of this system would allow smaller and more efficient reactors to be built which, by allowing the size of the submarine to be reduced, would lead to an increase in maximum speed. Without going through Rickover, representatives of the Defense Department set out to interview university scientists about the possibilities of new designs for lightweight nuclear reactors for submarines. When Rickover was notified of the initiative, he used congressional muscle to prevent the investigations. It seems that no method, no insight, no proposal into nuclear design or technology was acceptable unless he or the men who worked directly for him initiated it. There was, it appeared to observers, the navy way and the Rickover way and by the early 1970s Rickover was running a private force that was totally beyond the control of even the most senior naval appointments.

By 1974 there were thirty-four Yankee class submarines in the Soviet fleet each carrying the same number of ballistic missiles as each of the forty-one ballistic-missile submarines in the American fleet. But Gorshkov was already planning to reveal the Yankee's successor. The Soviet Union was developing the first generation of a class of submarines the West would come to know as the Delta. It was the largest ever built up to that time, displacing 10 200 tons when submerged, and although it was armed with only twelve missiles each was capable of hitting a target 4210 nautical miles (7600 km) away.

Then Gorshkov made two moves which ended any hope the United States might have had of regaining the lead in the underwater chess game. From the mid 1970s, the Soviet Union began to launch attack submarines that incorporated some of the most advanced submarine technology in the world. It was immediately designated the Alfa. The name was appropriate. The Soviet engineers had designated a liquid metal heat exchange system for the craft's nuclear reactor – the principle that Rickover, ten years before, had refused to reconsider as the way forward for his nuclear reactors. Instead of the pressurized-water reactor used on all previous Soviet nuclear submarines, Soviet engineers had come up with a lead-bismuth liquid metal coolant system. It was an incredible breakthrough because it made possible a smaller craft that could generate a speed of 43 knots, considerably faster than any American submarine. The Alfa's speed was the result of her hydrodynamic configuration, which helped to reduce acoustic noise levels. The shape was first committed to paper in the late 1950s by the team Peregoudov had recruited to work on the K-3.

The Alfa could dive to a depth of 2500 (760 m) with a safety margin of a further 1500 (450 m). Such a depth was possible because the pressure hull was made of titanium, a metal far stronger and lighter than steel. It is also non-magnetic, which made the craft safe from an entire range of anti-submarine mines. The workings of the submarine were highly automated and the crew, virtually all officers and warrant officers, was reduced probably to fewer than forty men, fewer than the number required to man other nuclear attack submarines. The performance of the Alfa contributed towards the US navy adapting its torpedo technology: the depths the Alfa could reach were too deep to allow unmodified American torpedoes to follow.

Gorshkov's second surprise, in the early 1980s, was another Soviet wonder that displaced 26 500 tons when submerged and was 80 ft (24 m) wide and more than 560 ft (170 m) long, with twenty missile tubes. In the West this class of submarine, once referred to as 'Taiphun' by Premier Leonid Brezhnev in a conversation with President Gerald Ford, came to be known as the Typhoon.

Even before the appearance of the Typhoon, Admiral Rickover said: 'The Soviets continue to make rapid progress in the design and construction of nuclear submarines in particular, as well as the rapid expansion and improved capabilities of their surface combatant fleet. As a student of naval history, I am concerned that there has never in peacetime been anything comparable to the current growth of Russian naval power'. But the Soviet momentum was far from spent.

In February 1984 Gorshkov watched the launch of the last generation of the Delta class of submarines. The Delta IV now had sixteen liquid propellant missiles of far greater accuracy, each of which had ten independent warheads capable of being programmed to strike anywhere within a range of 4500 nautical miles (8300 km).

The Delta IV complemented the Oscar class of guided-missile submarines which could travel at 28 knots and carry nuclear anti-ship missiles and anti-submarine missiles. Even then the Soviets had not finished. Just before Gorshkov retired in 1985 his third generation of submarines was nearing completion. When first launched in 1984, the Akula class, the newest Soviet attack submarine, had Western intelligence experts in a state of consternation. Although they knew the Soviets had reduced the noise levels of successive classes of submarines, making them much more difficult to detect, the Americans had felt confident of maintaining their lead in sophisticated acoustics. However, the Akula class confounded intelligence reports that the Soviets still lagged behind them. United States submarines tracking the first of the Akula class on sea trials soon after her completion, found her operating with unexpectedly low levels of noise, a development that experts had believed was still perhaps ten years away.

American recognition of the true state of relative technologies was inevitable. In a speech in 1986 Admiral Lee Baggett Jr, Commander-in-Chief of US forces in the Atlantic, said the American lead in submarine acoustics had narrowed to only three to four years at the outside. The following year, John Lehman, Secretary of the US navy, admitted that the Soviets had closed the gap and that their new submarines were virtually as quiet as the craft the Americans had built a few years before. Anthony Batista, another expert, claimed that the Akula class was not only a quieter craft than its predecessors but also the best submarine in the world.

USS *Salt Lake City*, one of the Los Angeles class submarines. Rickover had argued that the rapidly increasing Soviet threat made such a high-speed attack submarine essential.

The Alfa class attack submarine, introduced by the Soviets in the 1970s. A breakthrough in coolant system technology and a titanium hull meant a smaller craft which could reach 43 knots and dive to depths which defied US standard torpedoes.

It was inevitable that the United States was about to lose the lead in quality as well as quantity and in 1988 a panel of strategic thinkers on anti-submarine warfare suggested that the US navy would, in effect, have to 'start over' in its approach to such warfare.

Future developments in submarine technology are likely to follow the same trend. Though the Cold War is now over, intense submarine research and

development programmes are still in place and a state of competition still exists. One American analyst believes that in the years to come the only advantages the US navy will enjoy over the Russians will be its passive sonar capability and the quality of its submarine crews. Russian experts expect to be superior in all other areas of submarine technology by the end of this century or the early part of the next, with vessels capable of speeds of up to 50 or even 60 knots in the short term and perhaps 100 knots in the longer term at depths of more than

OVERLEAF USS *Billfish* undergoing hull blasting operations on a rainy night at a floating dock in New London, Connecticut.

HIPPINGPORT

ARDM-4

Gorshkov's Akula class attack submarine, first launched in 1984. US tracking submarines discovered unexpectedly low levels of acoustic noise. One American expert claimed she was the best submarine in the world.

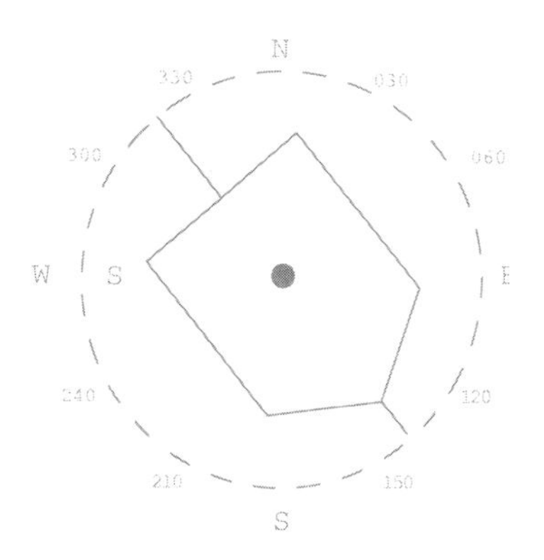

6000 ft (1830 m). Such performances will be made possible by technological advances which allow submarines to take in water through special ducts and expel it at great force astern to propel the craft through the water. Such a change from traditionally noisy screw propellers would further reduce the acoustic profile of the next generation of Russian submarines, which could well be armed with laser-guided torpedoes capable of speeds of up to 300 knots.

Other areas of research may have an even greater impact than modifications

to the propulsion unit. The United States and British navies are already employing pump-jet propulsion submarines, while refined hull shapes and frictionless coatings for hulls may play an equally important role in increasing the speed of submarines.

From the late 1950s up to the present day, what one observer has called 'the finite constraints' imposed upon United States submarine designers by Admiral Rickover have contrasted sharply with the Soviet system of four submarine-design bureaux which have kept abreast of all technological advances in other countries. Gorshkov's submarine design cycle led inexorably to Soviet developments that far exceeded Khrushchev's demand all those years ago to catch up with and surpass the Americans in nuclear submarine technology and performance.

Admiral Rickover, two years past his eightieth birthday, was eventually retired by the US navy in 1982. Admiral Gorshkov followed him three years later. Until the last days of the Cold War the submarine commanders they had chosen and trained waited for the signal that would trigger their part in the destruction of the world. In the end, despite American innovation and Soviet determination, there were no winners and no losers. Although hundreds of lethal attack, guided-missile and ballistic-bearing shapes have moved silently and secretly around the world at great depths since the Second World War, there has only been one order to attack. And that was a command from 8000 miles (12 900 km) away for a British nuclear submarine off the Falkland Islands to fire its conventional, high explosive torpedoes and sink the Argentinian light cruiser, *General Belgrano* – which was actually steaming away from the battle zone at the moment it was hit.

It was a last, familiar and deadly strike by a powerful remnant of the once dominant and unassailable British navy, determined to safeguard one of the few remaining rocky outposts of its former glorious empire on which the sun had never set.

After two hundred years of striving to produce a submarine boat which they could use to strike with certainty at the might of Great Britain's Royal Navy, any note of irony contained within the dying echo of that last explosion would not have been lost on the generations of inventors who lived and died with that dream as their simple inspiration.

USS *Jefferson City* surges through the Atlantic at her top surface speed in 1992 when she was the newest submarine in the fleet.

Notes on Sources

Many of the experiences in this book were related first-hand by ex-submariners and experts from all parts of the world and I would like to take this opportunity to say how grateful I am for their help and co-operation.

I would also like to express my thanks to the publishers and copyright holders of the following books for their permission to quote them. The pages on which the extracts appear are given at the end of each entry. Full bibliographical details are given in the Select Bibliography.

Prologue

Nautilus 90 North, W. R. Anderson. Reproduced by permission of Hodder & Stoughton (11–13)

Chapter 1 Catching Up

La Dramatique Histoire des Sous-Marins Nucleaires Sovietiques, L. Giltsov, L. N. Mormoul and L. Ossipenko (15, 19)

Chapter 3 The Race to be First

John P. Holland: The Inventor of the Modern Submarine, R. K. Morris. © 1966 by the US Naval Institute, Annapolis, Maryland, USA (43, 44, 45, 47, 48, 51–2, 58, 89–90)

'British Submarine Policy', M. Dash (47, 52–3, 59)

Building the Kaiser's Navy, G. E. Weir (61, 62)

Chapter 4 The Ultimate Test

U-Boat Stories, C. Neureuther and C. Bergen

Raiders of the Deep, Lowell Thomas. Copyright 1928 by Doubleday, a division of Bantam, Doubleday, Dell Publishing Group, Inc. Used by permission of Doubleday, a division of Bantam, Doubleday, Dell Publishing Group, Inc. (65–6, 67, 68, 85, 88, 98, 100)

Submarines and the War at Sea 1914–1918, R. Compton-Hall (82)

Max Horton and the Western Approaches, W. S. Chalmers (72)

From the Dreadnought to Scapa Flow, A. J. Marder (89)

U-Boats Destroyed, R. M. Grant (90–1, 96, 101)

Chapter 5 Challenge of the Deep

Half Mile Down, W. Beebe (103, 104, 105, 106, 107, 110, 113)

In Balloon and Bathyscaphe, A. Piccard (111–12)

Voyage of the Valdiva, Karl Chun (112)

Chapter 8 To the Bottom of the Sea

Seven Miles Down, J. Piccard and R. Dietz. Reprinted by permission of Curtis Brown Ltd, Copyright © 1961 by Jacques Piccard and Robert Dietz (175, 176, 177, 178, 179)

In Balloon and Bathyscaphe, A. Piccard (172, 174)

The Bombs of Palomares, T. Szulc (183)

'Captain Hook's Hunt for the H-Bomb', Marvin J. McCamis (184, 185–6, 189–90, 191)

Water Baby: The Story of Alvin, V. A. Kaharl (185, 186–9)

Chapter 9 Enter the Robots

The Discovery of the Titanic, R. Ballard. Reprinted by permission of Hodder & Stoughton and Warner Books, New York. Copyright © 1987 by Ballard and Family (200, 201)

Chapter 10 Masters of Inner Space

Rickover: Controversy and Genius, N. Polmar and T. B. Allen (211, 217, 219, 223)

Select Bibliography

ANDERSON, W. A. *Nautilus 90 North* Nelson, 1964; New American Library, 1959; Hodder & Stoughton, 1959

BALLARD, R. *The Discovery of the Titanic* Hodder & Stoughton, 1987

BEEBE, W. *Half Mile Down* John Lane, 1935

BURGESS, R. F. *Ships Beneath the Sea* Robert Hale, 1976

CHALMERS, W. S. *Max Horton and the Western Approaches*, Hodder & Stoughton, 1954

CHUN, KARL, *Voyage of the Valdiva*, privately published, 1901

COMPTON-HALL, R. *Submarine Boats* Windward, 1983 and Conway Maritime, 1983; *Submarines and the War at Sea* Macmillan, 1991

CRANE, J. *Submarine* BBC Books, 1984

DASH, M. 'British Submarine Policy' (unpublished thesis), 1990

DE MASSON, H. *Du Nautilus au Redoubtable* Press de la Cité, Paris, 1968

DUNCAN, F. *Rickover and the Nuclear Navy* US Naval Institute Press, 1990

GILTSOV, L., MORMOUL, N. and OSSIPENKO, L. *La Dramatique Histoire des Sous-Marins Nucleaires Sovietiques* Laffont, 1992

GRANT, R. M. *U-Boats Destroyed* Putnam, 1964

GRAY, E. *A Damned Un-English Weapon* Seeley, 1971; New English Library, 1973

KAHARL, V. A. *Water Baby: The Story of Alvin* Oxford University Press, 1990

MARDER, A. J. *From the Dreadnought to Scapa Flow* Oxford University Press, 1965

McCAMIS, MARVIN J. 'Captain Hook's Hunt for the H-Bomb', *Oceanus*, Vol. 31, No. 4, Winter 1989/90 (Woods Hole Oceanographic Institute)

MORRIS, R. K. *John P. Holland: The Inventor of the Modern Submarine* Arno Press, New York, 1980, 1966

NEUREUTHER, C. and BERGEN, C. *U-Boat Stories* Constable, 1931

PICCARD, A. *In Balloon and Bathyscaphe* Cassell and Macmillan Publishing Co, 1956

PICCARD, J. and DIETZ, R. *Seven Miles Down* Putnam, 1961; Longmans, 1962

POLMAR, N. *The American Submarine* Stephens, Cambridge, 1981 and US Nautical & Aviation Pub. Co., 1983; *Guide to the Soviet Navy* US Naval Institute Press, 1991

POLMAR, N. and ALLEN, T.B. *Rickover: Controversy and Genius* Simon & Schuster, 1982

POLMAR, N. AND NOOT, J. *Submarines of the Russian and Soviet Navies, 1718–1990* US Naval Institute Press, 1991

SILVERBERG, R. *The World of the Ocean Depths* World's Work, 1970

SWEENEY, J. B. *Pictorial History of Oceanographic Submersibles* Robert Hale, 1972

SZULC, T. *The Bombs of Palomares* Gollancz, 1967

THOMAS, L. *Raiders of the Deep* Heinemann and Doubleday, 1928

TYLER, P. *Running Critical* Harper & Row, 1986

WEIR, G. E. *Building American Submarines, 1914–1940* US Naval Historical Center, Department of the Navy, 1991; *Building The Kaiser's Navy: the Imperial Naval Office and German Industry in the von Tirpitz era, 1890–1919* US Naval Institute Press, 1992

Conversion Chart

Conversions of imperial and metric measurements of distance have been rounded up or down as follows:

Imperial	Metric
6 in	15 cm
1 ft	30 cm
2 ft	60 cm
4 ft	1.2 m
5 ft	1.5 m
10 ft	3 m
20 ft	6 m
30 ft	9 m
40 ft	12 m
50 ft	15 m
60 ft	18 m
70 ft	21 m
80 ft	25 m
90 ft	28 m
100 ft	30 m
200 ft	60 m
300 ft	90 m
400 ft	120 m
500 ft	150 m
600 ft	180 m
700 ft	215 m
800 ft	240 m
900 ft	275 m
1000 ft	305 m
2000 ft	610 m
3000 ft	910 m
4000 ft	1220 m
5000 ft	1500 m
10 000 ft	3000 m
20 000 ft	6000 m
30 000 ft	9000 m
35 000 ft	10 500 m

A number of measurements are peculiar to the sea:
1 fathom = 6 ft (1.8 m)
1 nautical mile = 1.15 statute miles (1.85 km)
A knot is a unit of speed equivalent to 1 nautical mile per hour

PICTURE CREDITS

BBC Books would like to thank the following for providing photographs and for permission to reproduce copyright material. While every effort has been made to trace and acknowledge all copyright holders, we would like to apologize should there have been any errors or omissions.

Archiv für Kunst und Geschichte 94–5, 99
Associated Press 183
Bibliothek für Zeitgeschichte Stuttgart 66, 93
Bildarchiv für Preusischer Kulturbesitz 79, 86–7
Bildarchiv für Preussischer Kulturbesitz/Hans Hubmann 134–5, 137, 142
Gino Birindelli 155, 157
Carina Dvorak 148 below
E. T. Archive 35, 38–9, 50–1, 71, 85, 160, 162
Mary Evans Picture Library 33, 36, 49, 54, 196 left
Christian Grube/Paterson Museum, New Jersey 46
Imperial War Museum 70, 83, 167, above
Imperial War Museum/Mrs Betty Hamilton 114
Teruyoshi Ishibashi 165
Mitsuma Itakura 167 below
Thomas Johnson 139 below
Frank Jordan 117 below
Yoshiteru Kubo 170
Captain Lennox Napier 117 above, 139 above
National Archives, Washington 210
New York Zoological Society/The Wildlife Conservation Society 109
Novosti 17, 208
Planet Earth Pictures 102, 198
Popperfoto 111, 113, 173, 175
Ian Potts 131
Salamander Books 10
Rurick Alexandrovitch Timoseeve 25, 30, 31
Eric Topp 124, 144
TRH Pictures 12, 42, 56–7, 60, 192, 216, 218, 224–5, 228–9
Ullstein Bilderdienst 118, 119
Maria Visintini 152, 154
Woods Hole Oceanographic Institution 194, 196 right, 202 above, 202–3
Captain Robert Worthington 128–9, 148 above
Yogi Inc/R. Y. Kauffman 215
Yogi Inc/Steve Kauffman 3
Yogi Inc 6, 14, 206, 222, 226–7, 230

INDEX

Figures in **bold** refer to pages with illustrations